Current Topics in Microbiology and Immunology

197

Editors

Gene Silencing in Higher Plants and Related Phenomena in Other Eukaryotes

Edited by P. Meyer

With 17 Figures

Springer-Verlag
Berlin Heidelberg New York
London Paris Tokyo
Hong Kong Barcelona
Budapest

Dr. Peter Meyer
Max-Delbrück-Laboratorium
in der Max-Planck-Gesellschaft
Carl-von-Linné-Weg 10
50829 Köln
Germany

Cover Illustration:
The front page shows three different examples of gene silencing phenomena in plants: (1) Silencing of a transgenic pigmentation marker in petunia flowers due to DNA methylation (background photo, provided by Iris Heidmann). See p. 15 for further details. (2) Inhibition of tomato fruit ripening by antisense technology (photos in the upper two panels, provided by Don Grierson). Wild-type tomatoes (right) and antisense transformants (left) are shown. See p. 77 for further details. (3) Silencing of anthocyanin pigmentation in maize anthers by paramutation at the Pl locus (photos in the lower two panels, provided by Garth Patterson). The Pl-Rh phenotype (right) and the Pl'-mah phenotype (left) are shown. See p. 121 for further details.

The photo on the back page shows examples of flower phenotypes derived from co-suppression of a gene of the pigmentation pathway in petunia (photo provided by Richard Flavell). See p. 43 for further details.

Cover design: Künkel + Lopka, Ilvesheim

ISSN 0070-217X
ISBN-13: 978-3-642-79147-5 e-ISBN-13: 978-3-642-79145-1
DOI: 10.1007/978-3-642-79145-1

Production: PRODUserv Springer Produktions-Gesellschaft, Berlin
Typesetting: Thomson Press (India) Ltd, New Delhi
SPIN: 10474049 27/3020-5 4 3 2 1 0 – Printed on acid-free paper.

Preface

In March 1993, Richard Jorgensen, Amar Klar and Rob Martienssen organized a conference at the Bunbury Center in Cold Spring Harbor, where different epigenetic phenomena in bacteria and eukaryotes were discussed. At the Bunbury conference some scientists who worked on plant silencing phenomena discussed the initiation of a common research project. The idea to discuss gene silencing phenomena in a volume of the CTMI series was born at a symposium in Cologne that had been organized to establish a European research network on plant gene silencing. At this symposium, the participants presented their work to the members of the Max-Planck-Institute for Breeding Research. After the talks, Heinz Saedler convinced the participants of the gene silencing network to present the current state of the art of the different silencing systems in a CTMI volume and to complement this summary with reviews on related phenomena in other eukaryotes.

The term 'gene silencing' refers to a complete or partial inactivation of gene activity. Silencing has been observed for endogenous plant genes and transposable elements, but most frequently for recombinant genes introduced into transgenic plants. In transgenic plants, gene silencing can occur either in primary transformants or during further breeding and propagation of transgenic plants. Transgenes or endogenous genes can also become silenced under the influence of a second homologous copy. The silencing effect can be unidirectional (trans-inactivation) or bidirectional (co-suppression) and it can be influenced by developmental and environmental factors. Inactivation is either associated with a loss of transcription, often corresponding to hypermethylation within the promoter region, or attributed to post-transcriptional degradation of RNA.

Certain silencing events show similarities with gene inactivation phenomena in other eukaryotes. In *Drosophila* the regulatory function of chromatin structure for gene expression has been convincingly documented. Mechanisms such as position effect variegation and transvection that are mediated by modifications in chromatin structure might serve as a reference model to discuss trans-inactivation or paramutation in plants. In filamentous fungi, homology-dependent inactivation mechanisms are present, reminiscent of trans-inactivation in plants. DNA methylation, which is involved in the regulation of transcription of plant transgenes and transposable elements, also has an important role in developmental regulation in mammalian cells. Despite these similarities, however, the regulation and biological function of the common molecular mechanisms probably differ between plants and other eukaryotes. Also, gene silencing effects based on post-transcriptional regulation appear to be specific to plants.

The identification of the underlying mechanisms for the different gene silencing phenomena in plants will not only improve the stability of gene expression in transgenic plants, a major prerequisite for the application of transgenic material. It should also provide a better understanding of cellular mechanisms that control promoter activity, RNA transport, RNA stability or other steps involved in gene expression.

The authors of the plant articles have tried to simplify the understanding of their very specialized research topics by providing some basic information on the common genetic and molecular tools used in plant molecular biology. I hope that this presentation especially encourages readers who are not working on plants to use this book to learn about a fascinating topic and to discover similarities with their own research systems. The four articles on gene silencing phenomena in *Drosophila*, filamentous fungi and mammalian systems should provide a basis for a comparative evaluation.

Considering the common tendency in biological research to work on more and more specialized subjects, this book might stimulate some interdisciplinary research projects to compare gene regulation mechanisms in different eukaryotes.

I am most grateful to the authors of this volume who provided excellent articles at short notice. I would like to thank Heinz Saedler for initiating this book, Sarah Grant and Bob Dietrich for helping me with the editing process, and Marga Botsch and Doris Walker for correspondence and technical editing.

PETER MEYER

List of Contents

List of Contributors

(Their addresses can be found at the beginning of their respective chapters.)

trans-Inactivation of Homologous Sequences in *Nicotiana tabacum*

A.J.M. MATZKE and M.A. MATZKE

1 Introduction

In this chapter, we focus on a specific class of homology-dependent gene silencing, epistatic *trans*-inactivation, in one plant species, *Nicotiana tabacum* (tobacco). Epistatic *trans*-inactivation is defined as a nonreciprocal interaction that occurs between homologous, or partially homologous, transgenes present at nonallelic (ectopic) chromosomal locations, and the outcome is that one transgene locus becomes inactivated in the presence of the second (reviewed in MATZKE et al. 1994b). This phenomenon possibly results from the unidirectional transmission of a hypermethylated state at one locus to a homologous unmethylated region at the second locus. There are now several examples of epistatic *trans*-inactivation in transgenic tobacco. We will review these cases and contrast them with other classes of homology-dependent gene silencing such as paramutation (an allelic interaction) and co-suppression/sense suppression (a *reciprocal* ectopic interaction). Even though the consequences of co-suppression and epistatic *trans*-inactivation are similar, i.e., silencing of homologous genes, they can be distinguished by a number of features, thus pointing toward fundamentally

Institute of Molecular Biology, Austrian Academy of Sciences, Billrothstrasse 11, 5020 Salzburg, Austria

different mechanisms for the two processes. In contrast, there are notable similarities between some paramutation systems and epistatic *trans*-inactivation.

2 Experimental System

Tobacco is a dicotyledonous plant that has been widely used in studies on plant gene expression, primarily because leaf cells can be transformed easily with foreign DNA and subsequently regenerated into whole, fertile plants. Indeed, this tractability is reflected in the fact that tobacco was one of the first plants to be engineered genetically (for example, see BARTON et al. 1983). Tobacco is a natural allotetraploid originating from two ancestral diploid species, *N. sylvestris* and either *N. tomentosiformis* or *N. otophora* (KENTON et al. 1993). Nevertheless, with respect to inheritance of foreign genes, tobacco behaves as a diploid.

A common method for transforming tobacco uses vectors based on the transferred DNA (T-DNA) region of the *Agrobacterium tumefaciens* tumor-inducing (Ti) plasmid in conjunction with the so-called leaf disc method. In this, tobacco leaf pieces are incubated with a suspension of *Agrobacterium* cells containing a modified Ti plasmid that includes an antibiotic selection marker able to function in plant cells. Single leaf cells that have received the selectable marker gene (and any other genes positioned between the borders of the T-DNA) are able to divide and differentiate on a medium containing the appropriate antibiotic and a plant hormone to induce shoot regeneration.

Although this procedure provides a simple way to produce transgenic tobacco plants, achieving expression of transferred genes has not been so straightforward. In any given experiment it was possible to find a subset of transformants that expressed strongly a given gene; nevertheless, there were almost invariably plants that exhibited low or unstable expression. Such cases were usually attributed to ill-defined position effects. However, with the recent realization that low (or no) expression was often coupled to the presence of multiple copies of homologous transgenes at both linked and unlinked chromosomal locations, a new framework has evolved for studying and understanding anomolous behavior of transgenes in plants.

3 Multiple Copies of Transgenes Are Often Poorly Expressed

That multiple copies of transgenes could negatively affect expression was only discovered when investigators focused on plants exhibiting weak activity of transgenes. This was not initially a priority, since early studies were mainly concerned with plants displaying strong and/or tissue-specific expression of transgenes. Several studies on T-DNA structure and activity in transgenic petunias revealed a loose association between low expression and multiple and/or rearranged T-DNA copies integrated at a single locus (JONES et al. 1987; DEROLES and GARDNER 1988).

The clearest indication that multiple copies of homologous transgenes could lead to silencing of transgene expression in tobacco came in sequential transformation experiments in which a previously active transgene locus, *K*, became inactive and methylated following the introduction—by retransformation—of a second, partially homologous construct, H (Fig. 1; MATZKE et al. 1989). An important aspect that distinguished these experiments from previous ones was that both the methylation and the inactivation were gradually reversible when the second ("silencing") transgene locus was segregated away from the first ("target") locus. This established that the silencing of the target locus *K* was directly dependent on the presence of the silencing locus *H*, and not just due to progressive inactivation and methylation, as would be expected of a foreign DNA response (BESTOR 1990; DOERFLER 1991). A second novel feature was that unlike previous studies involving multiple transgene copies at a single locus, the target locus and silencing locus were clearly shown to be unlinked. This introduced the notion that copies of homologous transgenes on nonhomologous chromosomes could somehow interact to modify gene expression and induce epigenetic alterations.

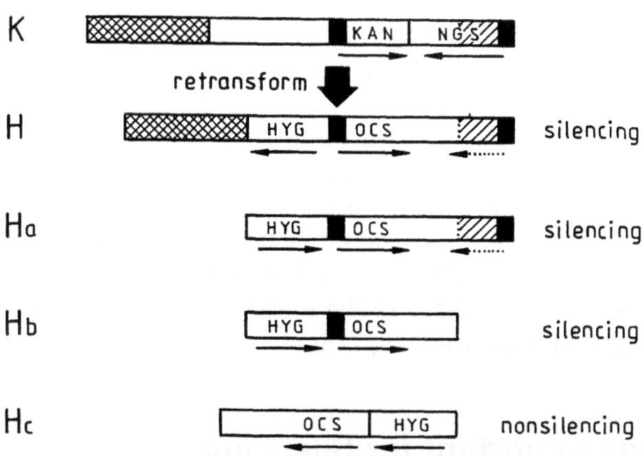

Fig. 1. Homology requirement for epistatic *trans*-inactivation in tobacco. A previously active transgene locus *K*, which encoded resistance to kanamycin (*KAN*) and nopaline synthase (*NOS*), became inactive and methylated when the *H* construct was introduced by retransformation (MATZKE et al. 1989) Regions of homology on *K* and *H* consisted of four copies of the NOSpro (*black boxes*), approximately 800 bp *nos* coding sequence (*slashed region*) and approximately 2.5 kb at the left border (*cross-hatched region*). Modified H constructs (*Ha, Hb, Hc*) were tested for their ability to *trans*-inactivate the target *K* locus. Homology in the NOSpro alone (*Hb*) was sufficient to silence and methylate both the KAN and NOS markers at the *K* locus (data not shown) An H construct in which all regions of homology to *K* were removed (*Hc*) did not silence *K* locus markers. The hygromycin resistance marker (*HYG*) was always under the control of the 35Spro, as was the OCS marker in Hc. *Arrows* show the orientation of the genes

Two other studies in transgenic tobacco confirmed the basic observation that homologous, or partially homologous, unlinked transgenes could interact in plant nuclei, leading to reduced expression of a target transgene locus. HOBBS and co-workers (1990) identified transgenic tobacco lines that either strongly or weakly expressed a transgene construct that encoded resistance to kanamycin and β-glucuronidase activity. Strongly expressing lines normally contained single, unmethylated copies of the construct, whereas weakly expressing lines contained more than one copy and these were methylated. When one weakly expressing line, containing two methylated copies of the construct arranged as an inverted repeat, was crossed with a strongly expressing line, the activity of the latter was reduced. This decreased activity was accompanied by increased methylation. Both the methylation and the inactivation were reversed when the interacting loci segregated in progeny (HOBBS et al. 1993).

VAUCHERET (1993) crossed a transgenic tobacco line that was homozygous for a silencing locus with more than 20 other tobacco lines that contained a variety of different target transgene constructs. The primary region of homology present between the silencing locus and various potential target loci consisted of the 35S promoter (35Spro) and/or the 19S promoter (19Spro) of cauliflower mosaic virus. Inactivation of all target transgenes was observed, even when the region of homology was reduced to 90 bp of the 35S promoter. In the silenced target transgenes examined, increased methylation was detected. The observed inactivation was at least partially reversible after crossing out the silencing locus. The silencing locus, which itself was not expressed, contained multiple, methylated copies of the transgene construct.

Three separate studies using transgenic tobacco plants thus established that a previously active, unmethylated target transgene locus could be reversibly inactivated and methylated when combined with a silencing locus that shared regions of homology with the target. Using these three studies as models for epistatic *trans*-inactivation, we will now consider the following points: (1) What regions of homology are important for epistatic silencing? (2) What are the characteristics of epistatic silencing loci? (3) Do potential target loci vary in their susceptibility to epistatic silencing? (4) What is the mechanism of epistatic silencing? (5) What are the genetic implications of the gradual recovery of target locus expression after segregating out a silencing locus?

4 Region of Homology Important for Silencing

Different regions of homology might be required to bring about either co-suppression or epistatic silencing. For example, in all cases of co-suppression, homology has comprised the protein coding region, and in only some cases, promoter segments in addition. Indeed, the usual design to achieve co-suppression is to place a cDNA copy of a gene under the control of a strong promoter, such as the 35Spro, and introduce the construct into the appropriate plant that has the endogenous homologous gene driven by its own promoter. Instead of overexpression of the gene, which might be expected with such constructs, expression of both the transgene and the homologous endogenous gene often collapses completely (reviewed in JORGENSEN 1990, 1992).

Although coding region homology effectively elicits co-suppression, several studies suggest that homology in promoter regions alone is sufficient to provoke epistatic *trans*-inactivation. In our original constructs, the target and silencing locus shared several regions of homology in addition to promoter fragments (MATZKE et al. 1989). Modified constructs have since established that homology comprising the nopaline synthase promoter (NOSpro) only is able to induce silencing of the target locus (Fig. 1). The silencing locus described by VAUCHERET (1993) also was homologous only to the promoter regions of inactivated target genes.

It can be provisionally concluded that the basic homology requirements for co-suppression and epistatic *trans*-inactivation differ: the former relies on coding sequence homology, whereas the latter can be observed when homology is limited to promoter regions. An unanswered question is whether DNA sequences that neither encode proteins nor supply traditional regulatory regions can be involved in homology-dependent gene silencing.

5 Characteristics of Epistatic Silencing Loci

In the study of HOBBS and co-workers (1990, 1993) and that of VAUCHERET (1993), each respective silencing locus was either weakly expressed or not expressed at all, and each silencing locus contained multiple copies of the transgene construct, which were methylated. We propose that this feature—hypermethylation—endows a locus with silencing ability. This proposal is derived from our comparative analysis of several transgene loci that contained the same construct but had variable silencing effects on a given target locus.

Three transgene loci (containing the same construct, H) that differentially affected a target transgene locus, *K*, were identified in a genetic analysis (MATZKE et al. 1993). In the presence of the H_1 locus, *K* was neither inactivated nor methylated. When combined with H_3, the *K* locus was partially inactivated and methylated when hemizygous (*K*-), and more completely inactivated and methylated when homozygous (*KK*). The H_2 locus induced complete inactivation and methylation of *K* in both the hemizygous and the homozygous state.

An analysis of the copy number of the H construct at the three *H* loci demonstrated that H_1 contained a single copy of the H construct, whereas H_2 and H_3 comprised multiple copies. Moreover, H_2 and H_3 were methylated in the region of homology common to the H construct and the inactivated genes at *K*, i.e., the

NOSpro. There was also a correspondence between the degree of methylation at H_2 or H_3 and the degree of methylation each imposed on *K*: H_3, which partially methylated *K*, was also partially methylated; H_2, which completely methylated *K*, was likewise completely methylated. In contrast, H_1, which did not induce methylation of *K*, was not methylated in the NOSpro. That the degree of methylation determined the silencing strength was further substantiated by generating epialleles of the strongly silencing H_2 locus: Two of these were less methylated and they no longer completely inactivated or methylated *K* in first-generation progeny (MATZKE et al. 1994a).

These results seemed to indicate that silencing loci contained multiple, methylated copies of the H transgene construct, and that the *degree* of methylation of a silencing locus could determine the extent to which it inactivated and methylated *K*. If an *H* locus contained an unmethylated copy of the H construct, as was the case with H_1, *K* remained unmethylated and continued to be expressed. (As a demonstration that partially homologous, unmethylated transgene copies can be expressed indefinitely, it is significant that a 6-year-old transgenic tobacco plant with a genotype of KK/H_1- still stably expresses both the *nos* and *ocs* genes, the biochemical markers at the *K* and *H* loci, respectively.) Although all epistatic silencing loci examined so far contain multiple linked copies of a given transgene construct [multiple copies tend to methylate de novo by an unknown mechanism (MITTELSTEN SCHEID et al. 1991; ASSAAD et al. 1993)], it should be noted that even single copies of transgenes can become methylated (PRÖLS and MEYER 1992) and these could conceivably act as silencing loci.

An important point relevant for the epistatic nature of the phenomenon is that the H_2 and H_3 silencing loci were methylated prior to interacting with *K*, whereas *K* only became methylated and inactivated after encountering H_2 or H_3 (MATZKE et al. 1993, 1994a). In other words, the hypermethylated state of a silencing locus determines its epistatic character: Approximately the same degree of methylation attained independently by a silencing locus can be imposed on the target *K* locus. This process is not accompanied by significant change in the methylation state of the silencing locus (Fig. 2). In contrast, *reciprocity* is characteristic of co-suppression: The expression of both the transgene and the endogenous gene is required to observe the ultimate effect, which is the reduction of steady state mRNA levels from both genes (KOOTER and MOL 1993; FRAY and GRIERSON 1993).

We have recently analyzed additional silencing loci containing the transgene constructs Ha and Hb (Fig. 1). In further support of our proposal, the ability to silence the target *K* locus was correlated with degree of methylation at a given *H* locus (A. Matzke and M. Matzke, unpublished data). Thus, all cases of epistatic silencing loci so far examined in tobacco appear to conform to the pattern of multiple, methylated transgene copies (we have not addressed the arrangement or completeness of these copies). This has implications for possible mechanisms (Sect. 7).

The methylated state of epistatic silencing loci is reminiscent of at least some examples of paramutation; a silencing locus would simply be analogous to a paramutagenic allele. Methylated paramutagenic alleles have been described for

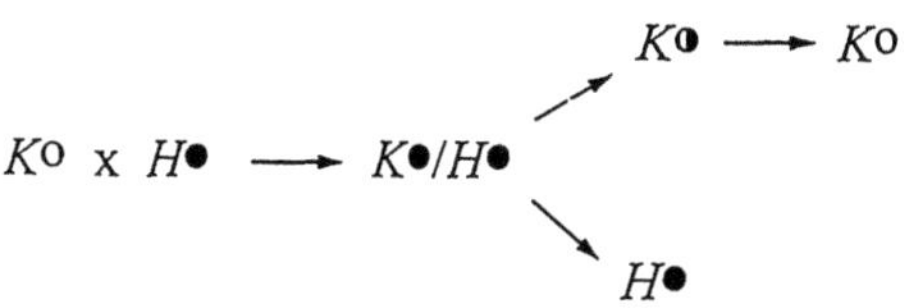

Fig. 2. Epistatic *trans*-inactivation. A silencing locus, *H*, independently acquires methylation ($H^{\bullet}$) because it contains multiple copies of the transgene construct H and/or because it has integrated into methylated plant DNA. When a susceptible, partially homologous target locus K° is combined with $H^{\bullet}$, it becomes inactivated and acquires a comparable degree of methylation ($K^{\bullet}$). This possibly occurs via a DNA pairing-dependent process of "epigene conversion" (MATZKE et al. 1994a). When $K^{\bullet}$ and $H^{\bullet}$ segregate, $K^{\bullet}$ only progressively reactivates and loses methylation ($K^{\bullet} \rightarrow K^{◑} \rightarrow K$). Thus, $H^{\bullet}$ heritably imprints K°, but does not itself change appreciably as a result of the interaction. *Symbols:* A *superscript open circle* denotes a susceptible target locus, i.e., one that has a slight inclination to spontaneously methylate, but normally only increases significantly in methylation after interacting with a silencing locus. A hypermethylated state is designed by a *superscript closed circle.* A *superscript half-closed circle* indicates partial loss of methylation. This means specifically the loss of methylation in some cells but not others, thus setting up a cellular mosaic of methylation at single restriction enzyme sites (MATZKE and MATZKE 1991; NEUHUBER et al. 1994)

a maize A_1 transgene locus in petunia (MEYER et al. 1993) and for the *R* locus in maize (J. Kermicle and M. Alleman, personal communication).

6 Susceptibility of Potential Target Transgene Loci to *trans*-Inactivation

The possibility that different (trans)genes could vary in their susceptibility to silencing effects originally came under consideration because of several reported cases in which introduction of a homologous transgene did not lead to any co-suppression involving the endogenous homologous gene. To account for these cases, it was suggested that (trans)genes could vary in their sensitivity to homology-dependent gene silencing (MATZKE and MATZKE 1993). This issue has become less clear-cut with the realization that different types of gene silencing are probably due to different mechanisms; therefore, the lack of silencing in a potential case of co-suppression may not be relevant for potential cases of epistatic *trans*-inactivation. Nevertheless, an analysis of transgene susceptibility to any type of homology-dependent gene silencing could reveal unsuspected features.

Such an investigation was carried out using four independent tobacco transformants, each containing the same transgene construct (H) integrated at different chromosomal locations (NEUHUBER et al. 1994). Susceptibility of each *H* locus to *trans*-inactivation was to be evaluated by retransforming leaf disks of each H line with a second, partially homologous construct, K (regions of homology consisted of the 35Spro). Numerous regenerated "double transformants," which contained the *H* locus and the K construct integrated at different sites, were screened for inactivation of *H* locus markers. Two sources of inactivation were identified. The first, which was independent of any *K* locus, resulted from the fact that two H lines were found to be intrinsically unstable when homozygous; this instability was manifested as inactivation of one or both *H* alleles in individual cells, leading to mosaic expression of *H* locus markers in leaves. As a consequence, double transformants regenerated from single leaf cells of these two homozygous lines often contained irreversibly inactivated *H* marker genes. In contrast, the other two H lines were stably expressed, meaning that all (or most) leaf cells expressed *H* locus markers, and this was unchanged in regenerated double transformants.

Concerning the original question of susceptibility to *trans*-inactivation in the presence of a partially homologous *K* locus, it was only with the two unstable H lines that cases of *K* locus-dependent *trans*-inactivation were found. This inactivation was reversed when *K* was segregated away from *H*, and was correlated with increased methylation of the *H* locus in the presence of a silencing *K* locus. Moreover, the one silencing *K* locus studied in detail comprised at least some duplications of the *K* construct and was partially methylated in the 35Spro (NEUHUBER et al. 1994). This system was thus similar to other cases of epistatic *trans*-inactivation described above.

From this analysis it was concluded that *H* loci that are susceptible to epistatic *trans*-inactivation are inherently unstable, with a tendency to become methylated and less active spontaneously, i.e., even in the absence of any silencing *K* locus. These intrinsic characteristics were particularly pronounced when the *H* locus was homozygous, suggesting that interactions between *H* alleles were contributing to the observed instability. The additional presence of a partially homologous *K* locus could also trigger conversion of a susceptible *H* locus to a less active and more methylated state. Because the behavior and methylation patterns of *H-/K-* genotypes were more similar to *HH/- -* genotypes than to *H-/- -* genotypes, it was concluded that a *K* locus could substitute for an *H* allele in what was otherwise an interaction between *H* alleles.

Although this conclusion was derived from a single study, the proposal that susceptible transgene loci are intrinsically unstable and prone to methylation may be applicable to other cases of epistatic *trans*-inactivation and paramutation. For example, it has been shown that a paramutable (i.e., sensitive) *R* allele is inherently unstable, with a propensity to spontaneously decrease in activity even in the absence of a paramutagenic *R* allele (BRINK et al. 1968). In addition, an intensively studied target transgene locus, *K*, was also shown to be partially methylated in one plant without ever having encountered a silencing locus (MATZKE et al. 1993). Thus a paramutagenic allele or silencing locus might only serve to potentiate an intrinsic instability of a paramutable allele or target locus. This conclusion was reached for paramutable *R* alleles some time ago (BRINK et al. 1968).

Finally, recent results from crosses of transgenic tobacco plants containing different epistatic silencing loci also suggest that susceptibility might indeed be due to an intrinsic instability and tendency to methylate: The 35Spro-driven hygromycin resistance marker on the H_2 locus that we have described (recall that

H_2 itself is a potent silencer of NOSpro-driven genes) was more sensitive than any other target tested to inactivation by the epistatic silencing locus characterized by VAUCHERET (1993), which silences target genes under the control of either the 35Spro or 19Spro (H. Vaucheret, personal communication). Interestingly, the copies of the 35Spro at the H_2 locus were shown to be partially methylated even prior to being crossed with the Vaucheret silencing locus (MATZKE et al. 1994a). Therefore, it is possible that the extreme sensitivity of H_2 to inactivation by the Vaucheret silencing locus is related to its inherent inclination to become methylated.

7 Mechanism of Epistatic *trans*-Inactivation

The mechanism of homology-dependent gene silencing has been pursued with great interest, particularly because different experimental systems have led researchers to suggest a variety of competing mechanisms (reviewed in JORGENSEN 1992; MATZKE and MATZKE 1993). By now, it is quite clear that several cases of co-suppression are the result of post-transcriptional events (DE CARVALHO et al. 1992; KOOTER and MOL 1993). Accordingly, transgene loci that elicit co-suppression are often transcribed at a high level (DE CARVALHO et al. 1992). This presumably produces abundant RNA that triggers turnover processes, resulting in disappearance of the RNA from the steady state pool.

In contrast to the relatively well-accepted role of post-transcriptional events in co-suppression, it is not clear that similar processes are operating in epistatic *trans*-inactivation. Any mechanism based on antisense RNA or RNA turnover was initially ruled out because promoter homology alone was apparently sufficient to induce silencing of target genes. Instead, competition for nondiffusible transcription factors was suggested (MATZKE and MATZKE 1990). Even homology that is restricted to promoters, however, can still contribute short 5'-untranslated regions to the transcript (ca. 35 bp in the case of NOSpro and ca. 15 bp in the case of the 35Spro). Either sense or antisense RNAs transcribed from these short 5' leaders could conceivably trigger turnover of RNAs that have this sequence in common, even if they have different coding regions. Therefore, unless transcription run-on experiments are performed, it is not possible to completely rule out that post-transcriptional processes also play a role in epistatic *trans*-inactivation.

There is only one published case of epistatic *trans*-inactivation in which nuclear run-on transcription has been performed and this seemed to indicate that silencing was due to transcriptional inactivation (NEUHUBER et al. 1994). In the well-defined system involving silencing of NOSpro-driven genes (MATZKE et al. 1989, 1993, 1994a), transcription from the NOSpro was weak and difficult to detect above background in run-on transcription experiments. However, even in this case, preliminary results suggest that the target transgenes are inactivated at the transcriptional level (Y.-D. Park and A.J.M. Matzke, unpublished results).

Several cases of paramutation have been shown to be due to transcriptional inactivation (MEYER et al. 1993; PATTERSON et al. 1993). The system described by MEYER and co-workers is most relevant for comparisons to epistatic silencing since it involves a methylated paramutagenic allele that inactivates and methylates a paramutable allele.

Even though extensive evidence for transcriptional inactivation in cases of epistatic *trans*-inactivation is lacking, we believe that the available information implies a specific mechanism. The key observation is that all epistatic silencing loci examined to date are methylated in the regions of homology to the target locus, prior to interacting with the target. It is difficult to envisage how a methylated transgene locus could produce sufficient amounts of RNA homologous to the target locus to initiate RNA turnover. As mentioned above, transgene loci inducing co-suppression seem to be transcribed at a high level (DE CARVALHO et al. 1992). In complete contrast, epistatic silencing loci, which are methylated, are often weakly expressed (HOBBS et al. 1993) and one (VAUCHERET 1993) is not expressed at all. Furthermore, an RNA turnover mechanism is inconsistent with our observation that reducing the degree of methylation within the H_2 silencing locus (which should, in principle, increase its expression), *weakened* the silencing ability of two H_2 epialleles (MATZKE et al. 1994a). Finally, with respect to competition for transcription factors, methylated promoters should exclude, not sequester, transcription factors.

If RNA turnover and sequestration of transcription factors can be ruled out, what possibilities remain? We believe that a process requiring pairing of homologous DNA sequences is most consistent with results obtained from systems exhibiting epistatic *trans*-inactivation. The plausibility of this suggestion derives from related phenomena in other organisms: Pairing of duplicated sequences in some filamentous fungi can lead to methylation and inactivation of the repeated genes, and pairing of some alleles in *Drosophila* can lead to gene inactivation and chromatin condensation (reviewed in MATZKE et al. 1994b). Epistatic *trans*-inactivation thus would require DNA sequence homology to allow pairing of interacting loci, as well as a silencing locus that is methylated prior to the pairing event. We have incorporated these features into a model in which the hypermethylated state of the silencing locus is imposed on an unmethylated homologous target locus via a process of "epigene conversion" [a term proposed by SABL and LAIRD (1992) to account for nonmendelian features of Huntington's disease in humans]. The pairing of methylated and unmethylated homologous sequences could produce a hemimethylated intermediate that is the preferred substrate of maintenance methylase (MATZKE et al. 1994a). This model can also be applied to cases of paramutation involving methylation. It has in fact been suggested that an epigenetic difference between paired paramutagenic and paramutable alleles triggers conversion of the latter to a less active, methylated state (MEYER et al. 1993).

So far, there is no direct evidence that pairing is involved in either epistatic *trans*-inactivation or paramutation. Pairing is comparatively easy to envision for paramutation, since alleles might be able to interact more readily than unlinked homologous sequences. Indeed, it is difficult to conceive how an inactive, methylated allele can impose this state on a normally active, unmethylated counterpart unless direct physical contact is involved (MEYER et al. 1993). Therefore, the demonstration that an ectopic interaction can produce effects similar to an allelic interaction (NEUHUBER et al. 1994) helps to extend the concept of pairing to both processes.

In summary, the mechanism of epistatic *trans*-inactivation has not been unequivocally established. However, the common characteristics of known epistatic silencing loci, most notably independently acquired hypermethylation, argue for a mechanism based on transmission of an inactive (or less active) state from the silencing locus to a normally active homologous target.

8 Genetic Implications of Delayed Recovery of Inactivated Target Loci

Regardless of the mechanism of epistatic *trans*-inactivation, a distinctive feature is that the genetic potential of the target locus is changed in a way that persists even after crossing out a silencing locus. As originally described for paramutation of *R* alleles in maize (BRINK et al. 1968), epistatic silencing of transgenes in tobacco leaves a heritable imprint—methylation has been identified so far—on the target locus, which subsequently takes several generations to lose methylation and recover full activity after being separated from the silencing locus (MATZKE and MATZKE 1991). In contrast, it has been claimed for several co-suppression systems that expression of an endogenous gene returns immediately to wide-type levels following transgene segregation (DOONER et al. 1991; FRAY and GRIERSON 1993). This behavior is compatible with the RNA turnover mechanism believed to be operating in co-suppression, since as soon as the excess RNA resulting from transgene transcription is no longer available to trigger turnover, the endogenous gene should be expressed normally. The imprint persisting at a target locus following an interaction with an epistatic silencing locus is inconsistent with the abrupt reversal of inactivation expected with an RNA turnover mechanism, but entirely compatible with epigene conversion: The methylated silencing locus imprints the homologous target locus with a comparable degree of methylation, and if the loss of methylation is inefficient, the imprint lingers even in the absence of the silencing locus.

That the genetic potential of a target locus can be changed (temporarily) by exposure to a silencing locus imparts a nonmendelian quality to epistatic *trans*-inactivation, as was previously recognized for paramutation (BRINK et al. 1968). Mendelian genetics states that segregating genes do not have a residual influence on each other. However, it is now clear that this principle can be violated, at least in plants, because of heritable epigenetic alterations brought about by allelic or ectopic interactions between homologous DNA sequences.

If methylation of a target locus declines steadily over several generations, but to different degrees in individual plants, then substantial epigenetic variation could be produced in a population. Indeed, sibling plants with the same transgenotype were found to have different degrees of transgene methylation and hence, they displayed a range of Kan^R phenotypes (MATZKE and MATZKE 1991). If these results can be generalized to endogenous plant genes, it is likely that epigenetic variation arising from both the immediate and the lingering effects of homology-dependent gene silencing can contribute substantially to phenomena such as incomplete penetrance and variable expressivity.

9 Conclusions

We have argued that epistatic *trans*-inactivation involves the imposition of methylation from a hypermethylated silencing locus onto a susceptible, hypomethylated target locus by a mechanism that requires pairing of homologous DNA sequences. This process is nonreciprocal because the silencing locus independently achieves a hypermethylated state whereas the target locus requires interaction with the silencing locus before it acquires significant methylation. Epistatic *trans*-inactivation is thus distinct from co-suppression, which depends on transcription of a transgene and the endogenous homologous gene (or a second transgene copy) to achieve gene silencing via an RNA turnover mechanism. Additional features that could distinguish co-suppression from epistatic *trans*-inactivation are compiled in Table 1. Since information on these specific points is still lacking from many silencing systems, this list can provide a framework for future studies on homology-dependent gene silencing. Epistatic *trans*-inactivation does share characteristics with several paramutation systems; these characteristics include the involvement of a hypermethylated silencing (paramutagenic) locus and the imprinting of heritable epigenetic modifications on the target (paramutable) locus.

Table 1. Features that might be used to distinguish co-suppression from epistatic *trans*-inactivation

	Co-suppression	Epistatic *trans*-inactivation
Homology	Coding region	Promoter
Epigenetic modifications of target locus?	Variable	Methylation
Recovery	Immediate	Gradual
Mechanism	RNA Turnover	Transcriptional inactivation
Silencing locus	Highly transcribed	Methylated

Acknowledgements. We thank the Austrian Fonds zur Förderung der wissenschaftlichen Forschung and the Bundesministerium für Wissenschaft und Forschung for financial support. We also thank H. Vaucheret for communicating unpublished results.

References

Assaad F, Tucker KL, Signer ER (1993) Epigenetic repeat-induced gene silencing (RIGS) in Arabidopsis. Plant Mol Biol 22: 1067–1085

Barton K, Binns A, Matzke AJM, Chilton M-D (1983) Regeneration of intact tobacco plants containing full length copies of genetically engineered T-DNA, and transmission of T-DNA to R1 progeny. Cell 32: 1033–1043

Bestor TH (1990) DNA methylation: evolution of a bacterial immune function into a regulator of gene expression and genome structure in higher eukaryotes. Philos Trans R Soc Lond [Biol] 326: 179–187

Brink RA, Styles ED, Axtell JD (1968) Paramutation: directed genetic change. Science 159: 161–170

de Carvalho F, Gheysen G, Kushnir S, Van Montagu M, Inzé D Castresana C (1992) Suppression of β-1,3-glucanase transgene expression in homozygous plants. EMBO J 11: 2595–2602

Deroles SC, Gardner RC (1988) Analysis of the T-DNA structure in a large number of transgenic petunias generated by Agrobacterium-mediated transformation. Plant Mol Biol 11: 365–377

Doerfler W (1991) Patterns of DNA methylation—evolutionary vestiges of foreign DNA inactivation as a host defense mechanism. Biol Chem Hoppe Seyler 372: 557–564

Dooner HK, Robbins TP, Jorgensen RA (1991) Genetic and developmental control of anthocyanin biosynthesis. Annu Rev Genet 25: 173–199

Fray RG, Grierson D (1993) Molecular genetics of tomato fruit ripening. Trends Genet 9: 438–443

Hobbs SLA, Kpodar P, DeLong CMO (1990) The effect of T-DNA copy number, position and methylation on reporter gene expression in tobacco transformants. Plant Mol Biol 15: 851–864

Hobbs SLA, Warkentin TD, DeLong CMO (1993) Transgene copy number can be positively or negatively associated with transgene expression. Plant Mol Biol 21: 17–26

Jones JDG, Gilbert DE, Grady KL, Jorgensen RA (1987) T-DNA structure and gene expression in petunia plants transfórmed by Agrobacterium tumefaciens C58 derivatives. Mol Gen Genet 207: 478–485

Jorgensen R (1990) Altered gene expression in plants due to *trans* interactions between homologous genes. Trends Biotechnol 8: 340–344

Jorgensen R (1992) Silencing of plant genes by homologous transgenes. Ag Biotech News Info 4: 265N–273N

Kenton A, Parokonny AS, Gleba YY, Bennett MD (1993) Characterization of the Nicotiana tabacum L. genome by molecular cytogenetics. Mol Gen Genet 240: 159–169

Kooter JM, Mol JNM (1993) Trans-inactivation of gene expression in plants. Curr Opin Biotechnol 4: 166–171

Matzke MA, Matzke AJM (1990) Gene interactions and epigenetic variation in transgenic plants. Dev Genet 11:214–223

Matzke MA, Matzke AJM (1991) Differential inactivation and methylation of a transgene in plants by two suppressor loci-containing homologous sequences. Plant Mol Biol 16: 821–830

Matzke MA, Matzke AJM (1993) Genomic imprinting in plants: parental effects and trans-inactivation phenomena. Annu Rev Plant Physiol Plant Mol Biol 44: 53–76

Matzke MA, Primig M, Trnovsky J, Matzke AJM (1989) Reversible methylation and inactivation of marker genes in sequentially transformed tobacco plants. EMBO J 8: 643–649

Matzke MA, Neuhuber F, Matzke AJM (1993) A variety of epistatic interactions can occur between partially homologous transgene loci brought together by sexual crossing. Mol Gen Genet 236: 379–386

Matzke AJM, Neuhuber F, Park Y-D, Ambros PF, Matzke MA (1994a) Homology-dependent gene silencing in transgenic plants: epistatic silencing loci contain multiple copies of methylated transgenes. Mol Gen Genet 244: 219–229

Matzke MA, Matzke AJM, Mittelsten Scheid O (1994b) Inactivation of repeated genes—DNA-DNA interaction? In: Paszkowski J (ed) Homologous recombination and gene silencing in plants. Kluwer, Dordrecht, p. 271–307

Meyer P, Heidmann I, Niedenhof I (1993) Differences in DNA methylation are associated with a paramutation phenomenon in transgenic petunia. Plant J 4: 89–100

Mittelsten Scheid O, Paszkowski J, Potrykus I (1991) Reversible inactivation of a transgene in Arabidopsis thaliana. Mol Gen Genet 228: 104–112

Neuhuber F, Park Y-D, Matzke AJM, Matzke MA (1994) Susceptibility of transgene loci to homology-dependent gene silencing. Mol Gen Genet 244: 230–241

Patterson G, Thorpe CJ, Chandler VL (1993) Paramutation, an allelic interaction, is associated with a stable and heritable reduction of transcription of the maize *b* regulatory gene. Genetics 135: 881–894

Pröls F, Meyer P (1992) The methylation patterns of chromosomal integration regions influence gene activity of transferred DNA in Petunia hybrida. Plant J 2: 465–475

Sabl JF, Laird CD (1992) Epigene conversion: a proposal with implications for gene mapping in humans. Am J Hum Genet 50: 1171–1177

Vaucheret H (1993) Identification of a general silencer for 19S and 35S promoters in a transgenic tobacco plant: 90 bp of homology in the promoter sequence are sufficient for trans-inactivation. C.R. Acad Sci Paris 316: 1471–1483

DNA Methylation and Transgene Silencing in *Petunia hybrida*

P. Meyer

1 Introduction

Changes in DNA methylation patterns are frequently associated with the modification of gene expression in mammalian and plant cells. This article summarizes data about transgene silencing accompanied by DNA hypermethylation in transgenic petunia plants carrying the *A1* gene from *Zea mays*. Introduction of the *A1* marker gene triggers the accumulation of brick-red pelargonidin pigments in floral cells of transgenic *Petunia hybrida*. Therefore, the system provides a model to study transcription of transgenes. Petunia plants are especially suitable for long-term studies on transgene stability because individual plants, regenerated from leaf tissue or single protoplasts, can be preserved in tissue culture or in the greenhouse over several years. Defined genotypes are readily propagated and amplified via stem cuttings or protoplast culture. New flowers continuously emerge from plants grown in the greenhouse, which allows a constant monitoring of transgene expression. Transcription instabilities are detected by changes in floral pigmentation patterns, and epigenetic variants of isogenic material can be

Max-Delbrück-Laboratorium in der Max-Planck-Gesellschaft, Carl-von-Linné Weg 10, 50829 Köln, Germany selected and propagated as individual plants, thus providing sufficient amounts of plant tissue for molecular analysis.

The system has proven to be particularly useful for the identification of multiple parameters that influence DNA methylation patterns (LINN et al. 1990; PRÖLS and MEYER 1992; MEYER et al. 1992b). Furthermore it has provided an insight into the influence of DNA methylation on homology-dependent gene silencing phenomena (MEYER et al. 1993, 1994). A comparison of DNA methylation patterns with their accompanying chromatin conformation suggested that DNA hypermethylation is associated with a local condensation of chromatin, which points to a mutual interaction of these two molecular mechanisms in the regulation of gene expression.

2 DNA Methylation in Plants

While in most vertebrates less than 10% of all cytosines are methylated (VANYUSHIN et al. 1970), plants contain up to 30% of m^5C residues (ADAMS and BURDON 1985). The high proportion of methylated C residues is mainly due to two facts. Angiosperm genomes contain a higher proportion of CpG dinucleotides, the common recognition sequence for DNA methylation in animals and plants, and plants provide additional target sequences for DNA methylation, as methylated C residues are located within CpG or CpNpG sequences (GRUENBAUM et al. 1981). The higher rate of CpG dinucleotides in plants compared to vertebrates is probably due to differential degrees of CpG depletion. According to the deamination theory, m^5C residues can undergo deamination to thymine, which leads to a depletion of CpG dinucleotides and an elevation of TpG and CpA dinucleotides (COULONDRE et al. 1978). Depletion levels are determined by comparing the numbers of observed CpG dinucleotides with the level expected from the base composition. In animals CpG dinucleotides are depleted to levels between 0.15% and 0.35% (JOSSE et al. 1961; RUSSELL et al. 1976), while dicot and monocot genomes only show a depletion to levels of 0.68% and 0.79%, respectively (GARDINER-GARDEN et al. 1992). It has been suggested that different depletion levels reflect a more efficient mismatch repair system in plants or a lower degree of DNA methylation in plant cells that contribute to the germ cells (GARDINER-GARDEN et al. 1992).

The degree of DNA methylation varies among individual plant species. *Arabidopsis thaliana* contains only 6.3% cytosine methylation, one-fourth of the methylation levels of most other angiosperms (LEUTWILER et al. 1984). This difference, however, does not reflect a reduced degree of DNA methylation efficiency, but the low proportion of highly methylated repetitive DNA in *A. thaliana*. CpG depletion rates in *A. thaliana* do not differ from those of other angiosperm species (GARDINER-GARDEN et al. 1992), suggesting that methylation of genes occurs to a similar extent in all angiosperms. Interestingly, CpWpG trinucleotides (W = A or T) are depleted to the same extent as GpWpC trinucleotides which are not methylatable by CpG or CpNpG methylation. This result suggests that either depletion of CpNpG sequences is not due to deamination of m^5C residues or that sequences other than CpG or CpNpG can also provide targets for DNA methylation.

It has been proposed that methylation of symmetrical sequences serves as a signal for a maintenance methylase that recognizes hemimethylated DNA produced upon DNA replication. The presence of methylated C residues in the opposite strand provides the information to methylate a C residue in the complementary sequence on the newly synthesized strand (HOLLIDAY and PUGH 1975). Hemimethylated DNA is the preferred substrate for maintenance methylase (GRUENBAUM et al. 1982), which supports the assumption that symmetrical DNA methylation patterns are transmitted in a semi-conservative fashion. It is unclear whether methyltransferase contains a substrate specificity for both CpG and CpNpG methylation, or whether different methyl-transferase enzymes exist in plants. While in mouse a single methyltransferase gene has been found, a small multigene family with homology to cytosine-5 methylase has been identified in *A. thaliana* (FINNEGAN and DENNIS 1993). It is therefore conceivable that plants contain several enzymes with different methylation specificity, but so far it is uncertain whether the members of this multigene family actually encode DNA methyltransferases.

In contrast to a mutation of the murine DNA methyltransferase that caused abnormal development and lethality of embryos (LI et al. 1992), DNA methylation mutants in *A. thaliana* do not affect plant development or viability. The *DDM1* mutant (VONGS et al. 1993) shows a reduction of m^5C levels by more than 70%, but the plant develops normally and exhibits no striking morphological phenotype. A *Neurospora crassa* mutant lacking a functional DNA methyltransferase is also viable, but frequently creates duplications of chromosomes or chromosomal parts (FOSS et al. 1993), suggesting a role for DNA methylation in the control of chromosomal behaviour. Apparently the regulation and function of cytosine methylation differ in vertebrates and plants. General statements about a common function for DNA methylation in eukaryotes can therefore be misleading. In plants, an inverse correlation between gene transcription and cytosine methylation has been observed for certain controlling elements (CHANDLER and WALBOT 1986; SCHWARZ and DENNIS 1986) and transgenes (AMASINO et al. 1984; VAN SLOGTEREN et al. 1984). Induction of several endogenous genes is also linked to a loss of cytosine methylation. The tissue-specific transcription of maize storage protein genes (SPENA et al. 1983; BIANCHI and VIOTTI 1988) and C4 photosynthesis genes (NGERNPRASIRTSIRI et al. 1989) correlates with hypomethylation of the genes in tissue where they are transcribed.

Although such examples clearly suggest a correlation between gene repression and DNA methylation, other reports can be found in the literature that do not detect any changes in DNA methylation patterns although gene activity is altered. Certainly not all gene silencing effects are based on a regulation of gene transcription by DNA methylation. It is most likely that DNA methylation is mainly involved in the regulation of promoter activities, but not in post-transcriptional regulation. If changes in gene expression are due to post-transcriptional regulation, promoter activity would probably not be impaired and no significant changes in DNA methylation should be detectable.

It is difficult to exclude the involvement of DNA methylation in changing promoter activity, because most DNA methylation studies have limited accuracy as they use isoschizomeres. Very often DNA methylation within a DNA region can be accurately monitored by measuring the methylation sensitivity of a restriction site located within this region. However, the state of DNA methylation at a restriction site might not always correspond to the degree of methylation of a neighbouring sequence that is involved in promoter regulation. Genomic sequencing analysis provides a more precise tool as the methylation state of every C residue can be analysed. Genomic sequencing of a 900-bp region upstream from the translation start codon of the maize alcohol dehydrogenase gene did not reveal any cytosine methylation although the gene is silenced (Nick et al. 1986). Apparently the *Adh1* gene provides an example for genes that are not regulated by DNA methylation. It cannot be excluded, however, that changes in DNA methylation further upstream of the promoter region have an influence on repression of the gene. Differences in DNA methylation do not necessarily occur within the gene or its promoter region. The cell-specific transcription of the *PEPCase* gene of C_4 plants, for example, corresponds to demethylation of a region located 3.3 kb upstream of the gene (Langdale et al. 1991).

As mentioned above, DNA methylation in plants might influence gene expression by a different mode of action than DNA methylation in vertebrates. We may consider that a downregulation of a promoter region cannot only be acquired by methylation of transcription factor binding sites. Hypermethylation within or near a promoter region might also induce changes in chromatin conformation that could impair accessibility of the promoter for cellular proteins.

3 The *A1* Marker System

The *A1* gene marker has been employed to analyse certain parameters that influence DNA methylation and to define the molecular mechanism by which DNA methylation influences gene transcription in plants. Transfer of the *A1* gene of *Zea mays* into the whitish flowering *Petunia hybrida* line RI01 leads to salmon-red flower pigmentation (Meyer et al. 1987). The *A1* gene encodes a dihydroflavonol reductase (DFR), an enzyme of the anthocyanin pigmentation pathway. Due to a genetic block in the anthocyanin pigmentation pathway (Stotz et al. 1985) the receptor line RI01 produces only very low amounts of cyanidin and delphinidin, the two pigments that determine floral pigmentation in petunia. The almost white flowers of RI01 accumulate dihydrokaempferol, a substrate of the maize DFR. In transgenic plants expressing the *A1* gene, dihydrokaempferol is converted into leucopelargonidin, which is further processed by endogenous enzymes into a red pelargonidin pigment. *A1* activity can be followed by analysing the occurrence and stability of salmon-red pigmentation in the flower of transgenic plants. The system has the advantage that expression can be monitored in individual petal cells and that no staining procedure is required. Transgenic petunia plants can be propagated over several years via cuttings. Due to the continuous production of new flowers, expression of the transgene can be monitored over longer periods and under variable conditions. The different inactivation phenomena that were observed for transgenic petunia plants carrying the maize *A1* gene were exclusively due to inactivation of transcription that was accompanied by hypermethylation within the promoter region. So far, we did not detect any evidence for post-transcriptional modifications. Analysis of A1 activity revealed several modes of transgene methylation (Fig. 1) that differ in timing and efficiency. Most phenomena described for other plants that are also accompanied by transgene methylation probably correspond to one of the four categories outlined in Fig. 1.

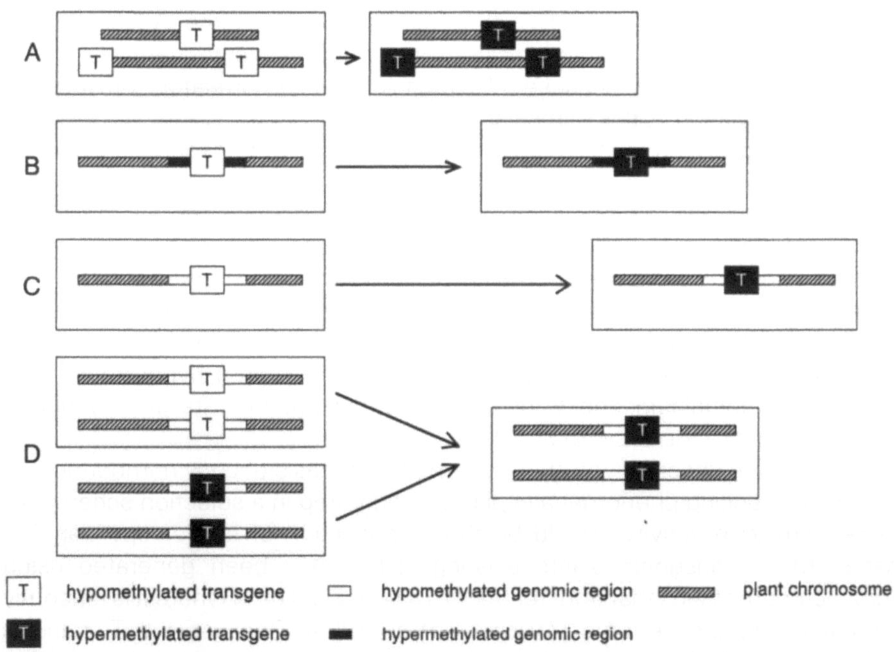

Fig. 1A–D. Different modes for the induction of *A1* methylation. **A** Multiple transgenes become efficiently and quickly methylated. **B** Upon integration into a hypermethylated genomic region a single-copy transgene becomes methylated due to a spread of the hypermethylation pattern. **C** A single-copy transgene that has integrated into a hypomethylated genomic region remains primarily hypomethylated. During prolonged propagation and especially under field conditions transgene-specific hypermethylation is induced, while the integration region remains hypomethylated. **D** Paramutation: The combination of the hypermethylated and the hypomethylated allele of a transgenic line induces hypermethylation in the previously hypomethylated allele

4 Copy Number-Dependent Gene Silencing

After direct DNA transfer of an *A1* gene linked to the viral CaMV 35S promoter into petunia protoplasts, primary transformants were regenerated that could be classified into three groups: plants with whitish flowers without any indication of *A1* activity, transformants with salmon-red flowers exhibiting continuous *A1* activity and plants carrying variegated flowers with *A1* activity present in some petal cells. The uniformity of petal pigmentation was inversely correlated with the number of integrated *A1* copies, as integration of multiple *A1* copies predominantly created transformants with a whitish or variegated phenotype (LINN et al. 1990). Methylation analysis using methylation-sensitive restriction enzymes detected hypermethylation within the promoter region of multiple-copy transformants. In most single-copy plants, however, a *Hpa*II restriction site at the boundary of the promoter and the *A1* cDNA remained unmethylated. This *Hpa*II site is located close to a 21-bp element containing two TGACG motifs that bind the activation sequence factor (ASF-1) required for the maximal expression of the promoter (LAM et al. 1989). Although the measurement of DNA methylation is limited to a few restriction sites, the *Hpa*II site next to the ASF-1 binding sequence could be used as an indicator for promoter-specific methylation that inhibited *A1* transcription, because *A1* expression always correlated with a lack of cytosine methylation at this particular site.

Other reports also confirm the inverse correlation of transgene activity and multiple transgene insertions in *Petunia hybrida*. The analysis of petunia plants transformed with the nopaline synthase gene and with a chimeric chlorophyll *a/b* binding protein gene, respectively, revealed that plants with high copy number complex insertions composed of multiple inverted repeats showed low levels of expression (JONES et al. 1987). In a petunia line carrying three copies of a *tms2* gene construct as a negative selectable marker, the expression of the transgene was also efficiently shut down and the sequence became hypermethylated, while plants containing one transgene copy did not lose its activity during vegetative culture (RENCKENS et al. 1992).

Despite the uncertainties about the molecular basis of copy number-dependent gene silencing phenomena in plants, a first step in a selection scheme for stable transgene activity should be the identification of single-copy transformants. Most transgenic plants existing today have been generated using *Agrobacterium* strains that transfer their T-DNA to plant cells where it is randomly integrated into the genome (ZAMBRYSKI et al. 1983). Because the T-strand is probably protected by bacterial proteins during its passage into the nucleus, T-DNA is frequently integrated without large internal deletions. Due to the mechanism of illegitimate integration (MAYERHOFER et al. 1991), however, T-DNA insertions are often rearranged, preferentially in the junction region between T-DNA and chromosomal DNA (SIMPSON et al. 1982). About one-third of T-DNA transformants carry insertions at only one locus (KONCZ et al. 1989). Depending on the bacterial strain (JORGENSEN et al. 1987) and on inoculation densities (JONES et al.

1987), multimers and inverted repeat structures are frequently observed when T-DNA is integrated at a single locus. Transformants derived from direct gene transfer into protoplasts generally contain truncated parts of the transferred plasmid, but relatively high ratios of single-copy integration events can be obtained when supercoiled DNA is used for transformation (KARTZKE et al. 1990). The choice of the proper transformation procedure and conditions will certainly influence the proportion of single-copy integration events that increase the chance for stable transgene expression.

5 Paramutation

In one transgenic line, we detected another copy number-dependent gene silencing effect, in which the combination of two homologous alleles that differed in their state of methylation resulted in a paramutation phenomenon (MEYER et al. 1993). The term "paramutation" describes a heritable change in gene function directed by an allele when expression of a paramutable allele is inhibited by a paramutagenic allelic homologue (BRINK 1956). When the hypermethylated *A1* allele of a single-copy transformant was crossed with a homologous but hypomethylated allele, the hypomethylated allele became methylated and silenced in a semi-dominant way. Most of the F1 plants that had received a hypomethylated *A1* allele from one parent and a hypermethylated allele from the other parent showed a significant reduction in *A1* expression. Flowers of these F1 plants were either white or highly variable with white and coloured spots or sectors within individual flowers. Again, inhibition of *A1* expression correlated with hypermethylation of the promoter region. Apparently the hypermethylated *A1* allele has paramutagenic potential, inducing methylation in the paramutable *A1* allele that had previously been hypomethylated and that remained hypomethylated in control crosses when it was not combined with the hypermethylated allele. After both *A1* alleles had segregated from each other, the paramutated *A1* allele remained hypermethylated over the next two generations. Only occasionally did a weak reactivation of the hypermethylated *A1* allele occur in a few cells.

Other groups also report an increase in transgene methylation for homozygous transgenes (MATZKE et al. 1993), but it is unclear whether the combination of two alleles in homozygous transformants generally leads to an increase in DNA methylation, or whether this effect is limited to certain loci. Interestingly, the induction of hypermethylation in tissue culture often results in homozygous changes with both alleles becoming methylated (KAEPPLER and PHILLIPS 1993). This observation supports the assumption that, at least at certain loci, methylation patterns can be exchanged among homologous alleles.

Most likely the paramutation-like behaviour of the *A1* allele is due to an interaction between the two differentially methylated alleles possibly mediated by transient pairing of the two alleles. Because we also observe a higher chromatin condensation in the hypermethylated allele (Ten Lohuis and Meyer, in preparation), we suggest that an exchange of methylation patterns also includes a change in chromosomal components. Assuming that transient pairing might not only occur between allelic homologues, but also between certain ectopic transgenes, some *trans*-inactivation phenomena might also be based on transient somatic pairing and an exchange of chromatin components.

6 Position Effects

Early studies on the activity of T-DNA genes in crown gall tumour lines revealed a correlation between DNA methylation and inactivity of T-DNA genes in certain lines (Amasino et al. 1984). Apparently T-DNA that integrates randomly into the genome becomes methylated at certain integration sites. The fact that certain copies of a transgene became hypermethylated while others remained hypomethylated and transcriptionally active, strongly suggests that the degree of DNA methylation is influenced by the integration region. Occasionally this position-specific inactivation was found to revert either spontaneously or after treatment of the cell line with the demethylating agent 5-azacytidine (van Slogteren et al. 1984; John and Amasino 1989).

To study position-dependent effects for de novo methylation of *A1* copies, we analysed single-copy transformants with different *A1* expression levels (Pröls and Meyer 1992). Although the integration of one transgene improves the chance for universal expression, it is not sufficient. Three transformants, lines 16, 17 and 24, were selected, each carrying one complete copy of the transgene integrated at different chromosomal loci. In the whitish line 16, no *A1* transcription was detectable; lines 17 and 24 showed *A1* transcription at high and low intensity, respectively. In line 16, the inactivated transgene had integrated into a highly repetitive and hypermethylated genomic region. The weakly expressed transgene in line 24 had integrated into a unique region that was highly methylated. The most intensively transcribed transgene in line 17 was inserted into a unique region that was hypomethylated. The characteristic hyper- and hypomethylation patterns of the integration regions in lines 24 and 17, respectively, were also imposed on the border region of the transgenes (Pröls and Meyer 1992). Apparently the methylation state of the integration region is, at least to some extent, responsible for position-dependent differences in gene expression.

The *A1* gene in line 17 remained hypomethylated and active in the majority of plants that were propagated in the greenhouse. Occasionally hypermethylation occurred in some flowers when plants were propagated over longer periods (Linn et al. 1990). Hypermethylation was significantly enhanced when line 17 was grown in the field (Meyer et al. 1992b), which allowed us to select epigenetic variants of line 17 containing the same transgene in different methylation states. A detailed analysis of the transgene and its chromosomal integration region in these epigenetic variants, but also in a derivative of line 17 that carries a deletion in the *A1* coding region, showed that hypermethylation was limited to the transgene DNA only, while the hypomethylation state of the integration region remained unaltered (MEYER and HEIDMANN 1994). This might suggest the presence of a DNA methylation mechanism that specifically recognizes foreign DNA. This mechanism, which has already been proposed for animal systems (BESTOR 1990; DOERFLER 1991), might identify foreign DNA by its nucleotide composition. It has been shown that the nuclear genomes of angiosperms are mosaics of long, compositionally homogeneous DNA segments called isochores (SALINAS et al. 1988), and that isochores contain very defined GC contents of functional genes and their chromosomal environment (MATASSI et al. 1989). The *A1* transgene in line 17 differs significantly in its AT content from the chromosomal environment. The integrated plasmid DNA has an average AT content of 47.5%, while the neighbouring 269 bp at the 5' end and 196 bp at the 3' end show a much higher average AT content of 74% and 77%, respectively. It is, therefore, conceivable that the isochore composition of a transgene has to match with that of its integration regions to avoid specific methylation of the transgene.

Alternatively it might not be the nucleotide composition but rather the transcriptional activity which makes the transgene a specific target for DNA methylation. Promoter activity might open the local chromatin structure, thus providing access in this region for cellular proteins involved in DNA methylation or heterochromatinization. So far, however, no data are available about DNA methylation of promoterless transgenes. For the future it will be interesting to compare DNA methylation patterns in transgenes embedded into different isochore regions and differing in nucleotide composition, secondary structure and promoter content.

As a first step, it will be important to define which characteristics of an integration region contribute to a stable expression. This would allow the selection of suitable integration regions into which transgenes might be targeted, if the presently very low efficiencies for gene targeting into the plant genome (PASZKOWSKI et al. 1988) can be improved. If a particular isochore composition of a transferred gene is important for its stable activity, expression might be improved when genes are transferred with suitable chromosomal regions they have been embedded in. It has been shown that the co-transfer of neighbouring chromosomal regions significantly stabilizes the expression of the human β-globin gene (GROSVELD et al. 1987). A similar result was obtained from work with the matrix attachment regions of the chicken lysozyme gene which co-map with the boundaries of the chromatin domain of this locus. A reporter gene that was flanked on both sides by a 3-kb attachment element from the 5′ region of the chicken lysozyme gene mediated elevated and position-independent transgene activity (STIEF et al. 1989). In plants similar stabilizing effects have been achieved with transgenes flanked by Scaffold attachment sequences (BREYNE et al. 1992; ALLEN et al. 1993; SCHÖFFL et al. 1993).

Another approach to uncouple transgene DNA from negative position effects could be its positioning on extrachromosomally replicating vectors (MEYER et al. 1992a), although this will probably impose restrictions on the size of foreign DNA that can be propagated in transgenic plants.

7 Endogenous and Environmental Factors That Influence DNA Methylation

About 30 000 isogenic *F1* plants that derived from backcrosses of the homozygous line 17 with RI01 were grown in the field. While blossoms on plants flowering early in the season were predominantly red, up to two-thirds of the later flowers on the same plants showed a reduction in *A1* expression. Again, the reduction in *A1* activity correlated with an increase in methylation. More than 95% of the greenhouse population of line 17, but only 37% of the field population, showed stable activity of the *A1* gene, apparently due to environmental effects specifically present in the field. Interestingly, an endogenous factor seemed to determine the susceptibility of plants towards this environmental stimulus. Progeny derived from early pollination of young parental plants was almost insensitive to inactivation, while in progeny from elder flowers which developed on the same parental plants, *A1* expression was considerably reduced (Meyer et al. 1992b). If an increasing degree of DNA methylation is imposed on the transgene with increasing age of the parental plant, such imprinted methylation patterns would be transferred to the progeny. Progeny from elder flowers would be more susceptible to environmental stimuli as they carry a higher methylation density already.

For practical breeding this model implies that breeders should use very young material for pollination to avoid selection of transgenes that might already have developed hypermethylation patterns. Transgene inactivation after environmental stimuli has also been observed in other systems, where tissue culture (Renckens et al. 1992) and heat stress (Walter et al. 1992) functioned as the environmental stimulus. Environmental stimuli also influence DNA methylation patterns and activity of mobile elements (Walbot 1988). Apparently environmental conditions induce a higher degree of flexibility of local DNA methylation patterns, which can result in inactivation of hypomethylated genes or reactivation of hypermethylated genes. Genes providing housekeeping functions are probably protected from variations in DNA methylation induced by environmental stimuli. For other genes, however, the modulation of local methylation and expression patterns as a response to changing environmental conditions might provide an advantage for fast evolutionary adaptations (Matzke and Matzke 1990).

8 Molecular Mechanism Inducing De Novo DNA Methylation

The molecular mechanisms that regulate de novo methylation are still not clearly defined, but high RNA levels or changes in chromatin structure might be involved in certain DNA methylation effects. When cDNA units of plant viroids were inserted into the genome of tobacco plants, methylation was specifically induced when the transformants carried high levels of viroid RNA (WASSENEGGER et al. 1994). This observation suggests that high mRNA levels can induce de novo methylation of the corresponding DNA, downregulating its own transcription. It remains to be seen whether RNA-mediated de novo methylation is a general phenomenon, or whether it requires extremely high transcript levels or RNA sequences that provide a certain secondary structure. At least for certain genes, RNA might be involved in the induction of de novo methylation.

Another mechanism for de novo methylation could be a change in chromatin conformation that serves as a signal for a cytosine methyltransferase system. A connection between DNA methylation and changes in chromatin structure in plants has been documented for rRNA genes in wheat (THOMPSON and FLAVELL 1988), where rRNA genes at different NOR loci are expressed at different levels. Nuclear dominance of particular NOR loci is associated with both hypomethylation and an increased sensitivity to DNase I digestion. Another indication for a regulation of DNA methylation by structural characteristics emerged from the genomic sequencing analysis of different methylation states of the *A1* transgene. Besides the common symmetrical methylation patterns, we observed methylation of cytosine residues located within non-symmetrical sequences (MEYER et al. 1994). Non-symmetrical m^5C residues were clustered, suggesting that some regions are more susceptible to methylation than others. In different cells, methylation was not always imposed on identical cytosine residues, but on cytosine residues located within a particular region. The information that leads to a heterogeneous distribution of non-symmetrical methylation patterns might therefore be coded by structural elements and not by sequence. In its hypermethylated state, the *A1* transgene forms chromatin that displays an increased resistance towards endo-nucleases (Ten Lohuis and Meyer, in preparation), suggesting that DNA methylation is accompanied by chromatin shielding. Our data show that both chromatin condensation and methylation of non-symmetrical recognition sites are involved in *A1* gene silencing. Both DNA methylation and chromatin condensation might interact to provide the molecular signals that regulate allelic or even ectopic *trans*-inactivation. It has been proposed that inactivation of homologous transgenes in plants might be mediated by an exchange of chromatin components during transient ectopic pairing among the homologous regions (JORGENSEN 1991). Up to now, no genomic sequencing data or nuclei sensitivity assays are available for most silencing systems in plants, but a detailed comparison of DNA methylation patterns and the accompanying chromatin structure should improve our knowledge about the regulation of gene silencing in plants.

9 Outlook

It remains to be seen whether other plant species show the same degree of variability in transgene expression and DNA methylation that we observe in our studies of *A1* activity in *Petunia hybrida*. It is most probable that different genetic backgrounds will have a decisive influence on the degree of genetic instability and on the sensitivity against environmental stimuli that modulate gene activity. Different species and even different tissues of a particular plant might vary in composition and specificity of cellular factors involved in DNA modification or chromatin conformation. A major point to be clarified in future research will therefore be the identification of chromosomal proteins involved in local chromatin condensation and their interaction with methylated DNA. Clearly a better understanding of the rules that regulate chromosome morphology is required before we can make predictions about long-term stability of transgenic DNA. To guarantee stable expression of transgenic constructs, molecular biologists probably must follow the same time-consuming selection procedures that are required for the generation of new lines in classical breeding programmes. The identification of the relevant structural and genetic parameters that regulate gene stability should not only speed up the process of stabilizing transgene activity, but also provide important data about the regulation of gene activity in plants.

Acknowledgements. The author wishes to thank Dr. Sarah Grant for critical reading of the manuscript. The work was supported by a grant from the Nielsen-Fond of the Max-Planck-Society.

References

Adams RLP, Burdon RH (1985) Molecular biology of DNA methylation. Springer, Berlin Heidelberg New York

Allen GC, Hall GE, Childs LC, Weissinger AK, Spiker S, Thompson WF (1993) Scaffold attachment regions increase reporter gene expression in stably transformed plant cells. Plant Cell 5: 603–613

Amasino RM, Powell ALT, Gordon MP (1984) Changes in T-DNA methylation and expression are asssociated with phenotypic variation and plant regeneration in a crown gall tumor line. Mol Gen Genet 197: 437–446

Bestor TH (1990) DNA methylation: evolution of a bacterial immune function into a regulator of gene expression and genome structure in higher eukaryotes. Philos Trans R Soc Lond [Biol] 326: 179–187

Bianchi MW, Viotti A (1988) DNA methylation and tissue-specific transcription of the storage protein genes of maize. Plant Mol Biol 11: 203–211

Breyne P, Van Montague M, Depicker A, Gheysen G (1992) Characterization of a plant scaffold attachment region in a DNA fragment that normalizes transgene expression in tobacco. Plant Cell 4: 463–471

Brink RA (1956) A genetic change associated with the R locus in maize which is directed and potentially reversible. Genetics 41: 872–889

Chandler VL, Walbot V (1986) DNA modification of a maize transposable element correlates with loss of activity. Proc Natl Acad Sci USA 83: 1767–1771

Coulondre C, Miller JH, Farabaugh PJ, Gilbert W (1978) Molecular basis of base substitution hotspots in Escherichia coli. Nature 274: 775–780

Doerfler W (1991) Patterns of DNA methylation–evolutionary vestiges of foreign DNA inactivation as a host defense mechanism. Biol Chem Hoppe Seyler 372: 557–564

Finnegan EJ, Dennis ES (1993) Isolation and identification by sequence homology of a putative cytosine methyl-transferase from Arabidopsis thaliana. Nucleic Acids Res 21: 2383–2388

Foss HM, Roberts CJ, Claeys KM, Selker EU (1993) Abnormal chromosome behaviour in Neurospora mutants defective in DNA methylation. Science 262: 1737–1741

Gardiner-Garden M, Sved JA, Frommer M (1992) Methylation sites in angiosperm genes. J Mol Evol 34: 219–230

Grosveld F, Blom van Assendelft G, Greaves DR, Kollias G (1987) Position-independent, high-level expression of the human beta-globin gene in transgenic mice. Cell 51: 975–985

Gruenbaum Y, Naveh-Many T, Cedar H, Razin A (1981) Sequence specificity of methylation in higher plant DNA. Nature 292: 860–862

Gruenbaum Y, Cedar H, Razin A (1982) Substrate and sequence specificity of a eukaryotic DNA methylase. Nature 295: 620–622

Holliday R, Pugh JE (1975) DNA modification mechanisms and gene activity during development. Science 187: 226–232

John MC, Amasino RM (1989) Extensive changes in DNA methylation patterns accompany activation of a silent T-DNA ipt gene in Agrobacterium tumefaciens-transformed plant cells. Mol Cell Biol 9: 4298–4303

Jones JDG, Gilbert DE, Grady KL, Jorgensen RA (1987) T-DNA structure and gene expression in petunia plants transformed by Agrobacterium tumefaciens C58 derivatives. Mol Gen Genet 207: 478–485

Jorgensen R (1991) Beyond antisense—how do transgenes interact with homologous plant genes? Tibtech 9: 266–267

Jorgensen R, Snyder C, Jones JDG (1987) T-DNA is organized predominantly in inverted repeat structures in plants transformed with Agrobacterium tumefaciens C58 derivatives. Mol Gen Genet 207: 471–477

Josse J, Kaiser AD, Kornberg A (1961) Enzymatic synthesis of deoxyribonucleic acid VIII. Frequencies of nearest neighbor base sequences in deoxyribonucleic acid. J Biol Chem 236: 864–875

Kaeppler SM, Phillips RL (1993) DNA methylation and tissue culture-induced variation in plants. In Vitro Cell Dev Biol 29P: 125–130

Kartzke S, Saedler H, Meyer P (1990) Molecular analysis of transgenic plants derived from transformation of protoplasts at various stages of the cell cycle. Plant Sci 67: 63–72

Koncz C, Martini N, Mayerhof R, Koncz-Kalman Z, Körber H, Redei GP, Schell J (1989) High-frequency T-DNA-mediated gene tagging in plants. Proc Natl Acad Sci USA 86: 8467–8471

Lam E, Benfey PN, Gilmartin PM, Fang R-X, Chua N-H (1989) Site-specific mutations alter in vitro factor binding and change promoter expression pattern in transgenic plants. Proc Natl Acad Sci USA 86: 7890–7894

Langdale JA, Taylor WC, Nelson T (1991) Cell-specific accumulation of maize phosphoenolpyruvate carboxylase is correlated with demethylation at a specific site >3kb upstream of the gene. Mol Gen Genet 225: 49–55

Leutwiler LS, Hough-Evans BR, Meyerowitz EM (1984) The DNA of Arabidopsis thaliana. Mol Gen Genet 194: 15–23

Li E, Bestor TH, Jaenisch R (1992) Targeted mutation of the DNA methyltransferase gene results in embryonic lethylity. Cell 69: 915–926

Linn F, Heidmann I, Saedler H, Meyer P (1990) Epigenetic changes in the expression of the maize A1 gene in Petunia hybrida: role of numbers of integrated gene copies and state of methylation. Mol Gen Genet 222: 329–336

Matassi G, Montero LM, Salinas J, Bernardi G (1989) The isochore organization and the compositional distribution of homologous coding sequences in the nuclear genome of plants. Nucleic Acids Res 17: 5273–5290

Matzke MA, Matzke EJM (1990) Gene interactions and epigenetic variation in transgenic plants. Dev Genet 11: 214–223

Matzke MA, Neuhuber F, Matzke AJM (1993) A variety of epistatic interactions can occur between partially homologous transgene loci brought together by sexual crossing. Mol Gen Genet 236: 379–386

Mayerhofer R, Koncz-Kalman Z, Nawrath C, Bakkeren G, Crameri A, Angelis K, Redei GP, Schell J, Hohn B, Koncz C (1991) T-DNA integration: a mode of illegitimate recombination in plants. EMBO J 10: 697–704

Meyer P, Heidmann I (1994) Epigenetic variants of a transgenic petunia line show hypermethylation in transgene-DNA: an indication for specific recognition of foreign DNA in transgenic plants. Mol Gen Genet 243: 390–399

Meyer P, Heidmann I, Forkmann G, Saedler H (1987) A new petunia flower colour generated by transformation of a mutant with a maize gene. Nature 330: 677–678

Meyer P, Heidmann I, Niedenhof I (1992a) The use of African cassava mosaic virus as a vector system for plants. Gene 110: 213–217

Meyer P, Linn F, Heidmann I, Meyer ZAH, Niedenhof I, Saedler H (1992 b) Endogenous and environmental factors influence 35S promoter methylation of a maize A1 gene construct in transgenic petunia and its colour phenotype. Mol Gen Genet 231: 345–352

Meyer P, Heidmann I, Niedenhof I (1993) Differences in DNA-methylation are associated with a paramutation phenomenon in transgenic petunia. Plant J 4: 89–100

Meyer P, Niedenhof I, Ten Lohuis M (1994) Evidence for cytosine methylation of non-symmetrical sequences in transgenic Petunia hybrida. EMBO J 13: 2084–2088

Ngernprasirtsiri J, Chollet R, Kobayashi H, Sugiyama T, Akazawa T (1989) DNA methylation and the differential expression of C_4. photosynthesis genes in mesophyll and bundle sheet cells of greening maize leaves. J Biol Chem 264: 8241–8248

Nick H, Bowen B, Ferl RJ, Gilbert W (1986) Detection of cytosine methylation in the maize alcohol dehydrogenase gene by genomic sequencing. Nature 319: 243–246

Paszkowski J, Baur M, Bogucki A, Potrykus I (1988) Gene targeting in plants. EMBO J 7: 4021–4026

Pröls F, Meyer P (1992) The methylation patterns of chromosomal integration regions influence gene activity of transferred DNA in Petunia hybrida. Plant J 2: 465–475

Renckens S, De Greve H, Van Montagu M, Hernalsteens J-P (1992) Petunia plants escape from negative selection against a transgene by silencing the foreign DNA via methylation. Mol Gen Genet 233: 53–64

Russell GJ, Walker PMB, Elton RA, Subak-Sharpe JH (1976) Double frequency analysis of fractionated vertebrate nuclear DNA. J Mol Biol 108: 1–23

Salinas J, Matassi G, Montero LM, Bernardi G (1988) Compositional compartmentalization and compositional patterns in the nuclear genomes of plants. Nucleic Acids Res 16: 4269–4285

Schöffl F, Schröder G, Kliem M, Rieping M (1993) An SAR sequence containing 395 bp fragment mediated enhanced, gene-dosage-correlated expression of a chimeric heat shock gene in transgenic tobacco plants. Transgen Res 2: 93–100

Schwarz D, Dennis E (1986) Transposase activity of the Ac controlling element in maize is regulated by its degree of methylation. Mol Gen Genet 205: 476–482

Simpson RB, O'Hara PJ, Kwok W, Montoya AL, Lichtenstein C, Gordon MP, Nester EW (1982) DNA from the A6S/2 crown gall tumor contains scrambled Ti-plasmid sequences near its junction with plant DNA. Cell 29: 1005–1014

Spena A, Viotti A, Pirrotta V (1983). Two adjacent genomic zein sequences: structure, organization and tissue-specific restriction pattern. J Mol Biol 169: 799–811

Stief A, Winter DM, Straetling WH, Sippel AE (1989). A nuclear DNA attachment element mediates elevated and position-independent gene activity. Nature 341: 343–345

Stotz G, de Vlaming P, Wiering H, Schramm AW, Forkmann G (1985) Genetic and biochemical studies on flavonoid 3'-hydroxylation in flowers of Petunia hybrida. Theor Appl Genet 70: 300–305

Thompson WF, Flavell RB (1988) DNase I sensitivity of ribosomal RNA genes in chromatin and nucleolar dominance in wheat. J Mol Biol 204: 535–548

Van Slogteren GMS, Hooykaas PJJ, Schilperoot RA (1984) Silent T-DNA genes in plant lines transformed by Agrobacterium tumefaciens are activated by grafting and 5-azacytidine treatment. Plant Mol Biol 3: 333–336

Vanyushin BF, Tkacheva SG, Belozersky AN (1970) Rare bases in animal DNA. Nature 225: 948–949

Vongs A, Kakutani T, Martienssen RA, Richards EJ (1993) Arabidopsis thaliana DNA methylation mutants. Science 260: 1926–1928

Walbot V (1988) Reactivation of the mutator transposable element system following gamma irradiation of seed. Mol Gen Genet 212: 259–264

Walter C, Broer I, Hillemann D, Pühler A (1992) High frequency, heat treatment-induced inactivation of the phosphothricin resistance gene in transgenic single cell suspension cultures of Medicago sativa. Mol Gen Genet 235: 189–196

Wassenegger M, Heimes S, Riedel L, Sänger HL (1994) RNA-directed de novo methylation of genomic sequences in plants. Cell 76: 567–576

Zambryski P, Joos H, Genetello C, Leemans J, Van Montagu M, Schell J (1983) Ti plasmid vectors for the introduction of DNA into plant cells without alteration of their normal regeneration capacity. EMBO J 2: 2143–2150

Transgene Inactivation in *Arabidopsis thaliana*

O. Mittelsten Scheid

1 Introduction

This review on transgene inactivation in *Arabidopsis thaliana* should not conceal the fact that—as in other plant species—there are numerous transformants which exhibit stable transgene expression and inheritance over many generations. Nevertheless, occasional loss of transgene expression in the progeny has been observed in many transformation experiments with *Arabidopsis*. In most cases, selection among the transgenic lines for those with a reliable gene expression is sufficient, and the exceptional lines can be treated as an experimental failure. However, the introduction of transgenes can be followed by silencing of foreign and endogenous genes, sometimes only in later generations, in attempts to achieve up- or downregulation of specific gene activity by overexpression or antisense inhibition as well as insertion mutagenesis. The interpretation of some results may therefore require consideration of the possibility of gene silencing.

Furthermore, and more importantly, the occurrence of gene silencing in

Friedrich Miescher Institute, P.O.Box 2543, 4002 Basel, Switzerland

Arabidopsis thaliana offers the chance to study an important and interesting biological phenomenon in a plant species whose qualification as a model organism in other respects is beyond dispute. The main attributes, such as the small size and space requirement of the plant, its short generation cycle and ample seed production, the low chromosome number and the existence of various ecotypes, were emphasised by LAIBACH as long ago as 1943. Since then, numerous publications have referred to further advantages, especially for plant molecular genetics. Dense genetic maps based on restriction fragment length polymorphisms (RFLPs), random amplified polymorphic DNAs (RAPDs), and morphological markers, transformation techniques and numerous mutations obtained by chemical and insertion mutagenesis are available, and a world-wide endeavour was made to characterise plant genes and their interaction in a genome with a size of only 100 Mb (for review see RÉDEI 1975, MEYEROWITZ 1989, KONCZ et al. 1992a). Although the model character does have its disadvantages (the amount of material and short lifetime of individual plants, the restricted tissue culture response, the tedious crossing procedure), there is no doubt that these are outweighed by the outstanding insights provided into the organisation of a plant genome. This contribution is intended to explain why this knowledge might turn out to be very helpful in understanding gene silencing in plants. I will therefore concentrate on the genetic perspective of transgene inactivation in *Arabidopsis*. For a more comprehensive discussion including biochemical and physiological aspects of possible underlying mechanisms the reader is referred to other contributions in this volume.

2 Unexpected Segregation of the Transgenic Phenotype

In the earliest reports on the introduction of marker genes into the genome of *Arabidopsis thaliana*, the analysis of genetic transmission was limited to three lines (LLOYD et al. 1986) or two lines (SHEIKHOLESLAM and WEEKS 1987). In both cases, the resistant phenotype segregated according to the expectation for single or multiple copies of the gene. The extension of genetic analysis to a larger number of independent transformants in later studies revealed that 10% of lines obtained after transformation with *Agrobacterium tumefaciens* exhibit reduced representation of the transgenic phenotype among their selfed progeny (13/124, FELDMANN and MARKS 1987; 2/20, SCHMIDT and WILLMITZER 1988; 1/10, VALVEKENS et al. 1988; 2/11, SANGWAN et al. 1991; 1/28, BOUCHEZ et al. 1993; 1/17, BECHTOLD et al. 1993). The frequency of this observation was similar for *Agrobacterium* infection of leaf disks (SCHMIDT and WILLMITZER 1988), roots (VALVEKENS et al. 1988), seeds (FELDMANN and MARKS 1987), zygotic embryos (SANGWAN et al. 1991) and whole plants (BOUCHEZ et al. 1993; BECHTOLD et al. 1993). The incidence of non-Mendelian segregation ratios in transformants obtained by direct gene transfer to protoplasts is even higher: figures of 25% (6/25, DAMM et al. 1989) and 50%

(14/28, Karesch et al. 1991) have been reported.

The deviation from the expected 75% transgene expressing progeny in the case of a single dominant gene ranged from 69% to the extreme of no resistant seedling. Of a total of 263 transformants analysed, 21 were reported to produce only 8%–69% resistant seedlings (Feldmann and Marks 1987; Schmidt and Willmitzer 1988; Valvekens et al. 1988; Damm et al. 1989; Karesch et al. 1991; Sangwan et al. 1991; Bechtold et al. 1993), and 19 transformants yielded 0%–4% resistant progeny (Feldmann and Marks 1987; Karesch et al. 1991; Bouchez et al. 1993). The observation of non-Mendelian segregation was made in a variety of ecotypes (Wassilewskija, Columbia, C24, Zürich) and for two selectable marker genes (*nptII*, *hpt*), regulated by different promoters (TR1', CaMV35S) (references as above).

In most cases, transformants with a deficiency in phenotype transmission have not been analysed for the presence of the insert in the original regenerant and in the sensitive progeny, thereby precluding conclusions about a correlation between the molecular structure of the transgenic DNA and the loss of expression. It should, however, be noted that an integration of multiple copies, often linked in one genetic locus, has been reported for nearly all transformation protocols.

Possible explanations for the under-representation of progeny expressing the transgene include physical instability due to deletion or mutation of the insert, the chimerical nature of transgenic plants, or a bias towards the wild-type segregants due to an insertion into essential genes. However, the following sections will illustrate that some cases, if not most, may have originated from an inconsistency between genotype and phenotype due to reduced expression of the transgene.

3 Analysis of Inactivated Transgenes

Silencing of transgenes has been observed in *Nicotiana tabacum*, *Petunia hybrida*, *Lycopersicon esculentum* and *Nicotiana sylvestris* (for review see other chapters in this volume) prior to their observation in *Arabidopsis*. As in the other species, all studies of inactive transgenes in *Arabidopsis* were initially based on incidental observations of lack of gene expression in the progeny of transgenic lines, similar to those described in Sect. 2. Common to all cases considered in this section is the evidence that gene expression is lost in spite of the presence of physically intact inserts which are transmitted in a silent state during somatic or germinal inheritance. Hence they represent cases of epigenetic inactivation. The following sections are intended to describe conformity and deviation, from various perspectives, among the data from several laboratories.

3.1 Type of Genes and Inactivation Frequency

Evidence for epigenetic inactivation has been reported for transformants with "foreign" genes, having no endogenous counterpart in the genome, as well as for transgenic plants with genes sharing homology with nuclear genes of *Arabidopsis*. The first group includes the selectable or visible marker genes hygromycin phosphotransferase (*hpt*, CHANG 1988, MITTELSTEN SCHEID et al. 1991, ASSAAD et al. 1993), neomycin phosphotransferase (*nptII*, KILBY et al. 1992, ASSAAD et al. 1993) and *rolB* (DEHIO and SCHELL 1994). These genes were fused to either the cauliflower mosaic virus 35S promoter (CHANG 1988; MITTELSTEN SCHEID et al. 1991; ASSAAD et al. 1993; DEHIO and SCHELL 1994) or the nopaline synthase promoter (KILBY et al. 1992). Transformation with genes of plant origin led to silencing in the case of alcohol dehydrogenase in a mutant complementation experiment (CHANG 1988), and to inactivation of chalcone synthase (DAVIES 1993). The chlorophyll *a/b* binding gene *cab140* and the tumour morphology shoots gene *tms2* were both inactivated in a line transgenic for the *tms2* gene fused to the *cab140* promoter (BRUSSLAN et al. 1993).

The fraction of transformants affected by gene silencing cannot be easily compared. If no further selection for silencing events was applied, values between 3% (DAVIES) and 50% (MITTELSTEN SCHEID et al. 1991) were reported. In other cases, silencing was found after mutagenesis in a selective procedure using a negative marker with a frequency of ca. 10^{-5} (BRUSSLAN et al. 1993; J. Brusslan and E. Tobin, personal communication).

In most described cases, gene silencing was observed only in the first generation after transformation, or even later. There was no experimental basis for an evaluation of whether gene silencing also affects gene expression during or directly after integration into the plant genome, since recovery of transformation events relied on the expression of selectable markers and transgenes silent from the beginning would have escaped detection. Gene silencing during somatic growth of the original transformant was seen for the *rolB* gene (DEHIO and SCHELL 1994).

3.2 Degree and Phenotype of Silencing

Transgene inactivation can alter the level of gene expression between independent transformants, in a spatial distribution within individual plants, or between individuals in a population of siblings.

The degree of gene silencing in individual plants and the resulting phenotype in the case of the non-cell autonomous resistance markers can vary and depends on both the conditions of selection and the level of residual gene activity. Some transgenic seedlings with inactive genes show growth inhibition and bleaching on selective medium to the same extent as the non-transformed wild type (MITTELSTEN SCHEID et al. 1991; KILBY et al. 1992). In other cases, they exhibit a broad spectrum of intermediate expression phenotypes (ASSAAD et al. 1993), a characteristic weak expression in certain lines (KILBY et al. 1992) or partial resistance in interaction between active and inactive genes (O. Mittelsten Scheid, unpublished, see Sect. 3.8). The inactivated *cab140* gene did not result in a visible phenotype and was scored by resistance of seedlings to naphthalene acetamide due to the co-ordinate inactivation of the *tms2* gene (Brusslan et al. 1993).

Lack of chalcone synthase expression has been found to result in a reduced accumulation of anthocyanin in plants grown under intensive light conditions (Davies 1993). Reduced *rolB* expression is apparent from a reversion to normal growth, with reversal of the multiple morphological changes induced by *rolB* (Dehio and Schell 1994). Both cell autonomous markers display visible sectors if gene expression is lost during somatic growth.

Among populations of progeny plants, silencing sometimes affects all individuals (Chang 1988; Mittelsten Scheid et al. 1991; Kilby et al. 1992) though the percentage of plants with silenced genes may be as low as 4%. This is strongly dependent on the character of the parental line, the genotype and the generation after transformation (Kilby et al. 1992; Assaad et al. 1993; Dehio and Schell 1994).

3.3 Structure and Stability of Inserts

In nearly every case, transgenic inserts in lines exhibiting inactivation consist of several molecules of the transforming DNA. Copy numbers range between 2 and ca. 10, in direct or inverted repeats, and sometimes with rearrangement of the DNA and complex border fragments (references as above). Multiple sequences alone are not sufficient for silencing (Kilby et al. 1992 described a line with four copies and stable expression), but lower copy numbers seem to decrease the probability of inactivation. This was evident in those lines of Assaad et al. (1993) in which the copy number had been reduced to one as a result of intrachromosomal recombination (Assaad and Signer 1992). Inactivation was rare in these lines relative to the parental line with several repeats. However, there was not always a simple correlation between the copy number of the promoter (one, two or three) and the frequency or degree of silencing (Assaad et al. 1993).

One plant obtained from a cross of lines carrying multiples of the inactive *hpt* gene was found to have a strongly reduced copy number, probably due to an intrachromosomal deletion event (Mittelsten Scheid et al. 1994). This was the only plant in which a previously silent gene after reactivation (see Sect. 3.7) was transmitted through meiosis. Restoration of functional and active genes from recombination traps inserted into the genome occurred with frequencies in the order of 10^{-4} to 10^{-6} (Peterhans et al. 1990; Assaad and Signer 1992). Whether inactivated transgenic inserts in general undergo such rearrangements with a high frequency is as yet unclear.

In one case, transgene inactivation was observed in the M2 population after mutagenesis treatment with ethylmethane sulphonate (Brusslan et al. 1993), known to induce mainly point mutations. It remains to be investigated whether the transition from the previously active to the inactivated state is indeed due to a change in DNA sequence or whether it reflects a different modification (J. Brusslan, personal communication).

Rearrangement of repetitive and inactivated genes in the ascomycete *Neurospora crassa* is accompanied by repeat induced point mutation (RIP; for review see SELKER 1990; chapter by Singer and Selker, this volume). There is no evidence for RIP-like mutations in *Arabidopsis*. Apart from the bulk rearrangements described above, silent inserts in *Arabidopsis* seem not to be subjected to rapid sequence mutations. Indirect evidence for this assertion is provided by the occasional return to the active state (see Sect. 3.7), with frequencies exceeding those expected for reversion mutations. Direct sequence analysis of several copies of stably silent transgenes after passage through three generations revealed a perfect sequence maintenance (MITTELSTEN SCHEID et al. 1994).

3.4 Inheritance and Genotype Dependence

Most of the lines carrying inactive transgenes transmitted the insert so that it was also inactive in the progeny (MITTELSTEN SCHEID et al. 1991; KILBY et al. 1992; ASSAAD et al. 1993; BRUSSLAN et al. 1993). In some cases, silencing even increased progressively over successive sexual, self-pollinated generations (KILBY et al. 1992; ASSAAD et al. 1993). Inheritance of the silent state even occurred if the seed-producing shoot was part of a sector within the parental plant in which the gene was silenced (DAVIES 1993). Sublines obtained from the same transformant after vegetative in vitro propagation varied in their inheritance of silenced genes (ASSAAD et al. 1993).

The assays for gene activity of the selectable markers were performed with seedlings, and silencing of selectable markers was apparent even at this developmental stage. Although gametes or very early embryonic stages were not analysed for gene activity, the silent state in these cases seemed to be inherited meiotically. In contrast, a previously silent *rolB* gene was found to be active in young seedlings (DEHIO and SCHELL 1994). The "resetting" of the silent *rolB* gene during gamete formation or fertilisation may indicate a specific mode of inactivation.

One genetic parameter with an obvious influence on the frequency of gene silencing is the number of transgene alleles. Although silencing was observed when the transgene was in hemizygous configuration, inactivation was more pronounced and more frequent in plants homozygous for the insert (ASSAAD et al. 1993; BRUSSLAN et al. 1993; DEHIO and SCHELL 1994).

3.5 Transcription

In all cases investigated, the level of transcript corresponding to the inactive gene has been found to be reduced (MITTELSTEN SCHEID et al. 1991; ASSAAD et al. 1993; BRUSSLAN et al. 1993; DAVIES 1993; DEHIO and SCHELL 1994). Transcription initiation measured in "run on" experiments was also reduced, although to a lesser extent than the steady state mRNA level (5x versus 25x, BRUSSLAN et al. 1993; 5x versus 100x, DEHIO and SCHELL 1994). These results suggest that silencing can involve both transcriptional and post-transcriptional regulation.

3.6 Methylation

Loss of gene expression is sometimes associated with DNA methylation (see the chapters by A.J.M. Matzke and M.A. Matzke, P. Meyer, N.V. Fedoroff, M.J. Singer and E.U. Selker, J.-L. Rossignol and G. Faugeron and W. Doerfler, this volume; for review of plant data see FINNEGAN et al. 1993). Although *Arabidopsis* nuclear DNA contains outstanding little ^{5m}C compared to other angiosperms (LEUTWILER et al. 1984; PRUITT and MEYEROWITZ 1986), it has been proposed that this DNA modification is correlated with regulation of endogenous genes (BURN et al. 1993). DNA analysis using methylation-sensitive restriction enzymes has provided evidence that some cytosine residues in transgenic inserts of *Arabidopsis* have been changed to ^{5m}C (CHANG 1988; MITTELSTEN SCHEID et al. 1991; KILBY et al. 1992; ASSAAD et al. 1993; BRUSSLAN et al. 1993; DAVIES 1993). Whether this methylation interferes with transcription, or whether non-transcribed sequences become methylated, remains an open question. Hypermethylation has been found to correlate with the inactive state of transgenes in some cases (ASSAAD et al. 1993; DAVIES 1993), and shown to increase with progressive silencing in successive generations (KILBY et al. 1992). Nevertheless, restriction patterns for the completely silenced transgenes have been found to be heterogeneous and erratic in the first generation after transformation (MITTELSTEN SCHEID et al. 1991), and more pronounced only in later generations (O. Mittelsten Scheid, unpublished), matching the observation of KILBY et al. (1992). Hypermethylation was found in only one of the silent lines of CHANG (1988). BRUSSLAN et al. (1993) reported that there were no differences in methylation pattern between their active and inactive transgenes. In addition, no methylation could be detected in the case of the inactive *rolB* gene (DEHIO and SCHELL 1994). Some transgenes have been found to carry slight methylation in expressing lines (KILBY et al. 1992; ASSAAD et al. 1993). It has to be considered that only those methylated cytosines located within appropriate restriction enzyme recognition sites are detected. The resolution of this method is therefore limited. In particular, the often used promoter of the cauliflower mosaic virus 35S transcript has only one suitable restriction site (LINN et al. 1990). It could be, therefore, that a causal correlation between hypermethylation and gene inactivity has escaped our notice so far. Backcrosses of a mutant, exhibiting strongly reduced methylation of repetitive DNA, with the wild type have allowed examination of the degree of methylation in subsequent generations. The results indicate that de novo methylation of the initially hypomethylated DNA in *Arabidopsis* may be a slow process (VONGS et al. 1993). The data for silenced transgenes available so far may be interpreted as indicating that methylation is a secondary effect of gene silencing rather than the initial modification turning off gene expression.

One argument for the involvement of DNA methylation in aberrant transgene expression in *Arabidopsis* stems from the observation that a previously silent *nptII* gene could be reactivated by application of the DNA methyltransferase inhibitor 5-azacytidine in tissue culture of root explants (KILBY et al. 1992). Azacytidine treatment has also been found to increase the frequency of kanamycin-resistant plants after *Agrobacterium* transformation, a finding interpreted as being due to enhanced gene expression from the selectable markers (Mandal et al. 1993). This treatment reduced the methylation of plant DNA flanking the T-DNA inserts (Mandal et al. 1994). Growth of *Arabidopsis* on azacytidine also resulted in demethylation of rDNA (Kilby et al. 1992). However, it has not yet been established whether the strongly reduced seed set and seedling viability seen in two later generations (Kilby et al. 1992) are due to inhibition of methylation or to other unknown effects. Any indirect effect of the inhibitor, e.g., a change in higher-order chromatin structure, could also influence gene expression.

3.7 Reversibility of Gene Inactivation

Some silent transgenes can become reactivated without 5-azacytidine treatment. Tissue culture of explants from whole plants has been found to reactivate gene expression to a limited extent [Mittelsten Scheid et al. 1991; Kilby et al. 1992 (in the latter study reactivation occurred in a few explants from one line in controls without 5-azacytidine treatment)]. Resistant seedlings may also arise spontaneously from some sensitive plants (E. Signer, personal communication), and it should be emphasised that the silent *rolB* gene generally becomes reactivated in each generation (Dehio and Schell 1994). Surprisingly, a silent *hpt* gene, stable when plants were self-pollinated, was reactivated in an outcross with the wild type or with a strain containing another, non-allelic silent insert (Mittelsten Scheid et al. 1991). Reactivation resulted in 6% fully resistant plantlets in reciprocal crosses. The frequency indicated that this reactivation, like inactivation, has a stochastic component. The reactivated state was labile: the complex transgenic insert again experienced silencing during transmission to the next generation, so that all resistant plants obtained in the crosses (apart from one individual with a DNA rearrangement, see Sect. 3.3) had sensitive progeny only (Mittelsten Scheid et al. 1991).

3.8 Genetic Effects of Inactive Transgenes

Repeats of transgenes within one genetic locus, silenced by *cis*-inactivation (Matzke et al. 1994), can interact with homologous genes in an ectopic position (*trans*-inactivation, Matzke et al. 1994) and display a "dominant" character. When a line containing silenced, linked repeats of the *hpt* gene was crossed with a homozygous line having an ectopic, active, single-copy insert, gene expression from the latter transgene seemed to be reduced in 10% of the progeny. The resulting phenotype ranged from full sensitivity to a wide range of partial resistance never seen in the parental lines (O. Mittelsten Scheid, unpublished). A similar "inactivation dominance" was described for crosses of plants with silent *chs* genes and *chs* overexpressing lines, with the F1 showing a high number of sectors (Davies 1993). The inactive *cab140* locus had an incomplete silencing effect on an expressing, different member of the *cab* gene family (Brusslan et al. 1993). In this case, the interacting loci were genetically linked. It has not been established to what extent sequence homology and/or genetic linkage was responsible for this effect.

4 Inactivation of Transposable Elements

Transposons, which represent a special class of transgenic DNA, have been reported to undergo inactivation in several plant species. This is reviewed in detail by N.V. Federoff (this volume). The *Arabidopsis* genome contains two transposon-like sequences which appear to have undergone transposition earlier in the evolutionary history of the species and to have been inactive since then (VOYTAS and AUSUBEL 1988; PELEMAN et al. 1991). No transposition of the heterologous transposon *Mu1* from maize transferred to the *Arabidopsis* genome could be detected over a period of three generations (ZHANG and SOMERVILLE 1987). In contrast, the endogenous transposon *Tag1* (TSAY et al. 1993) and two other heterologous transposable elements of maize origin were mobile in the *Arabidopsis* genome. The autonomous *Ac* and its non-autonomous *Ds* derivatives, as well as the *En-1* element, showed somatic and germinal excision from surrounding marker genes and reinsertion into other positions (VAN SLUYS et al. 1987; MASTERSON et al. 1989; SCHMIDT and WILLMITZER 1989; CARDON et al. 1993). Both *Ac* and *En-1* remained active for at least three generations after transformation (SCHMIDT and WILLMITZER 1989; CARDON et al. 1993). The frequency of excision was increased with the number of *Ac* transposase gene copies (DEAN et al. 1992; KELLER et al. 1992; BANCROFT and DEAN 1993) and the level of transposase mRNA (SWINBURNE et al. 1992). Many factors, such as position of the insert in the genome, promoter strength, splicing efficiency and the availability of host factors, may influence excision frequencies. There is no direct evidence that transposons are inactivated in *Arabidopsis* (FEDOROFF and SMITH 1993), and the *Ac* inserts in plants with high and low transposition activity do not appear to differ in their pattern of DNA methylation (KELLER et al. 1992). However, not all transformants containing complete *Ac* elements show excision, and individual sibling plants with a common genotype, carrying the same insert with a complete and an incomplete *Ac* containing T-DNA, exhibit highly variable excision frequencies in their progeny populations (DEAN et al. 1992). Plants homozygous for an *Ac* transgene show higher variegation rates than do hemizygotes, but not the same degree of variation between each other (DEAN et al. 1992). The *tms2* gene from *Agrobacterium tumefaciens*, conferring NAM sensitivity, seemed to become progressively inactivated during prolonged tissue culture in the case of a multiple T-DNA insertion when it was used to counterselect against plants carrying the neighbouring transposase gene (FEDOROFF and SMITH 1993). Germinal excision rates were found to be reduced in later generations (BANCROFT and DEAN 1993), although this could not be distinguished form environmental influences. Some of these features resemble data described for other transgenes in Sect. 3, and it is likely that epigenetic mechanisms may also contribute to the regulation of transposon activity in *Arabidopsis*.

5 Conclusions and Perspectives

With the growing number of transformation experiments in *Arabidopsis thaliana* in recent years, increasing evidence has been obtained to indicate that gene transfer and integration alone are not sufficient to ensure display of the expected phenotype. Although it is in general an exceptional occurrence, a variety of foreign and endogenous transgenes are inactivated, to various extents, in individual plants, sibling lines and independent transformants. Loss of expression may involve transcriptional and post-transcriptional regulation. Gene inactivation affects duplicated copies of genes within a single locus or in an allelic or ectopic position to each other. Gene silencing in *Arabidopsis* is reversible, though the tendency to lose expression remains. The silent state can be heritable and progressive, and may suppress homologous, active genes. Specific DNA methylation has been found to be associated with silent genes but its role as a primary event in silencing remains to be established.

Different concepts and models about how transgenes in plants may be inactivated are discussed in other contributions to this volume. At the moment, none of them can provide a satisfactory explanation for all the experimental data, but they will hopefully induce and stimulate further research. The study of inactivated transgenes in *Arabidopsis* holds promise of progress especially in the systematic analysis of the genetic basis of gene silencing, which is more difficult to perform in other species.

Independent lines containing similar copy numbers of transgenes are not always affected by silencing (Kilby et al. 1992). This has been attributed to their different positions of integration within the genome (for review: Matzke et al. 1994). Little is known as to why transgene inactivation depends on the genomic neighbourhood. Mapping active and inactive inserts, isolating their flanking plant sequences and specifying their influence on gene expression and silencing may provide valuable information in this respect. The growing number of transformants with characterised T-DNA or transposon insertions (for reviews see Koncz et al. 1992b, Feldmann 1992, Coupland 1992) probably includes interesting material for such approaches.

Investigation of the role of methylation in gene silencing should also be facilitated by knowledge about molecular genetics in *Arabidopsis*. Mutations reducing the methylation of several genomic sequences have been described (Vongs et al. 1993), and it will be interesting to see whether gene silencing still occurs in these lines. However, genetic combination of lines carrying inactive resistance markers and the hypomethylated mutants was found not to result in reactivation in two subsequent generations (E. Signer, personal communication). Similar crosses are under investigation to establish whether the methylation mutants have an influence on lines differing in their response to vernalisation, for which a correlation with methylation exists (BURN et al. 1993, Burn, personal communication). The identification of a gene that encodes a putative methyltransferase from *Arabidopsis* (FINNEGAN and DENNIS 1993) will permit attempts at overexpression and the study of gene silencing in plants with an elevated level of DNA methylation, provided that such an effect is not deleterious for the plant. All these experiments might yield information not only about the role of methylation itself but also about a possible involvement of this DNA modification in chromatin structure and subsequent effects on gene expression.

One promising approach is the identification of mutants affecting the incidence or maintenance of silencing. Attempts to screen for such mutants have been made in lines with a suitable genetic background. Two loci have been reported to act as enhancers of silencing in the case of the *rolB* gene (DEHIO and SCHELL 1994). A putative mutant that could not maintain the silent state of a transgene was found after mutagenesis in a line with a strongly silenced hygromycin resistance gene, showing a high level of transgene expression without rearrangement of the transgene (O. Mittelsten Scheid, unpublished). Further characterisation of these and other mutants is required. Whether gene silencing observed for transgenes is also relevant for endogenous genes may become apparent if such mutants display any morphological, physiological or developmental aberrations.

Further, gene silencing might also be a useful phenomenon for investigation of the function of specific plant genes. At present, site-directed "knock out" mutations of nuclear plant genes are not practicable, since gene targeting by homologous recombination has only been achieved at low frequencies in plants (PASZKOWSKI et al. 1988; LEE et al. 1990; OFFRINGA et al. 1990; HALFTER et al. 1992). A specifically reduced expression of many genes has been achieved by transformation with antisense constructs (for review: VAN DER KROL et al. 1988). However, it has been reported that inhibition of the *AKR* gene (*Arabidopsis* ankyrin repeat) is obtained only in transgenic plants with multiple copies of the antisense construct, although all transformants contained a similar level of antisense mRNA (ZHANG et al. 1992). It was stated by the authors that "current hypotheses on antisense regulation might not explain this result" (ZHANG et al. 1992). In addition, some *Arabidopsis* plants transgenic for sense constructs with a truncated gene also have a reduced level of transcripts from the endogenous gene, and their phenotype is very similar to plants with antisense constructs. An alternative interpretation of these data might be that the endogenous gene is inactivated whereas the expression of the transgene is not affected (unilateral *trans*-inactivation, MATZKE et al. 1994). If this is the case, gene silencing in *Arabidopsis* could supply a powerful method for specifically knocking out expression of other genes, provided that the introduction of additional, ectopic copies causes inactivation of resident genes at a reasonable frequency. The occurrence of the corresponding mutant phenotypes after transformation with sense constructs of several regulatory genes (*cal, ap1, ag, ap3*, M. Yanofsky, personal communication) represents promising evidence for such a suppressing effect of homologous transgenes.

In summary, instability of transgene expression in *Arabidopsis* resembles gene silencing in other plants in many respects. More than 50 years ago Friedrich Laibach stated at the end of a publication: "From this short review of our many years of experiments with *Arabidopsis thaliana* (L.) Heynh., it follows that this is clearly a very suitable subject for the investigation of genetics and developmental physiology, which in some respects rivals *Drosophila*, the show-piece of geneticists." (Laibach 1943, translated). I hope that this conclusion will also prove true for research on epigenetics; if it does, our basic understanding of gene expression and interaction in plants and our ability to apply this knowledge in biotechnology will be substantially improved.

Acknowledgements. Several colleagues have contributed to this article by communicating manuscripts, data prior to publication and descriptions of ongoing research: I want to thank Judy Brusslan, Joanne Burn, Gareth Davies, Christoph Dehio, Nigel Kilby, Csaba Koncz, Abul Mandal, Elliot Meyerowitz, Eric Richards, Jozef Schell, Ethan Signer, Elaine Tobin and Martin Yanofsky for their interest and support. I apologise to those whose work might not have been considered due to my lack of awareness. Further, I am grateful for valuable comments on the manuscript by María José Diéguez-Ceriani, Jean Masson, Frederick Meins and Jerzy Paszkowski. Thanks are also due to Karin Afsar for proof-reading. Experimental work in our laboratory was financially supported by the Swiss National Foundation grant 31-30118.90.

References

Assaad FF, Signer ER (1992) Somatic and germinal recombination of a direct repeat in Arabidopsis. Genetics 132: 553–566

Assaad FF, Tucker KL, Signer ER (1993) Epigenetic repeat-induced gene silencing (RIGS) in Arabidopsis. Plant Mol Biol 22: 1067–1085

Bancroft I, Dean C (1993) Factors affecting the excision frequency of the maize transposable element Ds in Arabidopsis thaliana. Mol Gen Genet 240: 65–72

Bechtold N, Ellis J, Pelletier G (1993) In planta Agrobacterium mediated gene transfer by infiltration of adult Arabidopsis thaliana plants. C R Acad Sci Paris Life Sci 316: 1194–1199

Bouchez D, Camilleri C, Caboche M (1993) A binary vector based on Basta resistance for in planta transformation of Arabidopsis thaliana. C R Acad Sci Paris Life Sci 316: 1188–1193

Brusslan JA, Karlin-Neumann GA, Huang L, Tobin EM (1993) An Arabidopsis mutant with a reduced level of cab140 RNA is a result of cosuppression. Plant Cell 5: 667–677

Burn JE, Bagnall DJ, Metzger JD, Dennis ES, Peacock WJ (1993) DNA methylation, vernalization, and the initiation of flowering. Proc Natl Acad Sci USA 90: 287–291

Cardon G, Frey M, Saedler H, Gierl A (1993) Mobility of the maize transposable element En/Spm in Arabidopsis thaliana. Plant J 3: 773–784

Chang C (1988) Molecular genetic studies in Arabidopsis thaliana. PhD thesis, California Institute of Technology, Pasadena, California

Coupland G (1992) Transposon tagging in Arabidopsis. In: Koncz C, Chua NH, Schell J (eds) Methods in Arabidopsis research. World Scientific, Singapore, pp 290–309

Damm B, Schmidt R, Willmitzer L (1989) Efficient transformation of Arabidopsis thaliana using direct gene transfer to protoplasts. Mol Gen Genet 217: 6–12

Davies G (1993) PhD thesis, University of Cambridge, UK

Dean C, Sjodin C, Page T, Jones J, Lister C (1992) Behaviour of the maize transposable element Ac in Arabidopsis thaliana. Plant J 2: 69–81

Dehio C, Schell J (1994) Identification of plant genetic loci involved in a post-transcriptional mechanism for meiotically reversible transgene silencing. Proc Natl Acad Sci USA 91: 5538–5542

Fedoroff NV, Smith DL (1993) A versatile system for detecting transposition in Arabidopsis. Plant J 3: 273–289

Feldmann KA (1992) T-DNA insertion mutagenesis in Arabidopsis: seed infection/transformation. In: Koncz C, Chua NH, Schell J (eds) Methods in Arabidopsis research. World Scientific, Singapore, pp 274–289

Feldmann KA, Marks MD (1987) Agrobacterium-mediated transformation of germinating seeds of Arabidopsis thaliana: a non-tissue culture approach. Mol Gen Genet 208: 1–9
Finnegan EJ, Dennis ES (1993) Isolation and identification by sequence homology of a putative cytosine methyltransferase from Arabidopsis thaliana. Nucleic Acid Res 21: 2383–2388
Finnegan EJ, Brettell RIS, Dennis ES (1993) The role of DNA methylation in the regulation of plant gene expression. In: Jost JP, Saluz HP (eds) DNA methylation: molecular biology and biological significance. Birkhäuser, Basel, pp 218–261
Halfter U, Morris PC, Willmitzer L (1992) Gene targeting in Arabidopsis thaliana. Mol Gen Genet 231: 186–193
Karesch H, Bilang R, Mittelsten Scheid O, Potrykus I (1991) Direct gene transfer to protoplasts of Arabidopsis thaliana. Plant Cell Rep 9: 571–574
Keller J, Lim E, James DW Jr, Dooner HK (1992) Germinal and somatic activity of the maize element Activator (Ac) in Arabidopsis. Genetics 131: 449–459
Kilby NJ, Leyser HMO, Furner IJ (1992) Promoter methylation and progressive transgene inactivation in Arabidopsis. Plant Mol Biol 20: 103–112
Koncz C, Chua NH, Schell J (eds) (1992a) Methods in Arabidopsis research. World Scientific, Singapure
Koncz C, Schell J, Rédei GP (1992b) T-DNA transformation and insertion mutagenesis. In: Koncz C, Chua NH, Schell J (eds) Methods in Arabidopsis research. World Scientific, Singapore, pp 224–273
Laibach F (1943) Arabidopsis thaliana (L) Heynh als Objekt für genetische und entwicklungsphysiologische Untersuchungen. Bot Arch 44: 439–455
Lee KY, Lund P, Lowe K, Dunsmuir P (1990) Homologous recombination in plant cells after Agrobacterium-mediated transformation. Plant Cell 2: 415–425
Leutwiler LS, Hough-Evans BR, Meyerowitz EM (1984) The DNA of Arabidopsis thaliana. Mol Gen Genet 194: 15–23
Linn F, Heidmánn I, Saedler H, Meyer P (1990) Epigenetic changes in the expression of the maize A1 gene in Petunia hybrida: role of numbers of integrated gene copies and state of methylation. Mol Gen Genet 222: 329–336
Lloyd AM, Barnason AR, Rogers SG, Byrne MC, Fraley RT, Horsch RB (1986) Transformation of Arabidopsis thaliana with Agrobacterium tumefaciens. Science 234: 464–466
Mandal A, Lång V, Orczyk W, Palva ET (1993) Improved efficiency for T-DNA-mediated transformation and plasmid rescue in Arabidopsis thaliana. Theor Appl Genet 86: 621–628
Mandal A, Sandgren M, Palva ET (1994) Inhibition of cytosine methylation allows efficient cloning of T-DNA tagged plant DNA of Arabidopsis thaliana by plasmid rescue. In Vitro Cell Dev Biol (in press)
Masterson RV, Furtek DB, Grevelding C, Schell J (1989) A maize Ds transposable element containing a dihydrofolate reductase gene transposes in Nicotiana tabacum and Arabidopsis thaliana. Mol Gen Genet 219: 461–466
Matzke M, Matzke AJM, Mittelsten Scheid O (1994) Inactivation of repeated genes—DNA-DNA interaction? In: Paszkowski J (ed) Homologous recombination and gene silencing in plants. Kluwer Academic, Dordrecht pp 271–307
Meyerowitz EM (1989) Arabidopsis, a useful weed. Cell 56: 263–269
Mittelsten Scheid O, Paszkowski J, Potrykus I (1991) Reversible inactivation of a transgene in Arabidopsis thaliana. Mol Gen Genet 228: 104–112
Mittelsten Scheid O, Afsar K, Paszkowski J (1994) Gene inactivation in Arabidopsis thaliana is not accompanied by an accumulation of repeat-induced point mutations. Mol Gen Genet 244: 325–330
Offringa R, de Groot MJA, Haagsman HJ, Does MP, van den Elzen PJM, Hooykaas PJJ (1990) Extrachromosomal homologous recombination and gene targeting in plant cells after Agrobacterium mediated transformation. EMBO J 9: 3077–3084
Paszkowski J, Baur M, Bogucki A, Potrykus I (1988) Gene targeting in plants. EMBO J 7: 4021–4026
Peleman J, Cottyn B, Van Camp W, Van Montagu M, Inzé D (1991) Transient occurrence of extrachromosomal DNA of an Arabidopsis thaliana transposon-like element, Tat1. Proc Natl Acad Sci USA 88: 3618–3622
Peterhans A, Schlüpmann H, Basse C, Paszkowski J (1990) Intrachromosomal recombination in plants. EMBO J 9: 3437–3445
Pruitt RE, Meyerowitz EM (1986) Characterization of the genome of Arabidopsis thaliana. J Mol Biol 187: 169–183
Rédei GP (1975) Arabidopsis as a genetic tool. Annu Rev Genet 9: 111–127
Sangwan RS, Bourgeois Y, Sangwan-Norreel BS (1991) Genetic transformation of Arabidopsis thaliana zygotic embryos and identification of critical parameters influencing transformation efficiency. Mol Gen Genet 230: 475–485
Schmidt R, Willmitzer L (1988) High efficiency Agrobacterium tumefaciens-mediated transformation of Arabidopsis thaliana leaf and cotyledon explants. Plant Cell Rep 7: 583–586

Schmidt R, Willmitzer L (1989) The maize autonomous element activator (Ac) shows a minimal germinal excision frequency of 0.2%–0.5% in transgenic Arabidopsis thaliana plants. Mol Gen Genet 220: 17–24

Selker EU (1990) Premeiotic instability of repeated sequences in Neurospora crassa. Annu Rev Genet 24: 579–613

Sheikholeslam SN, Weeks DP (1987) Acetosyringone promotes high efficiency transformation of Arabidopsis thaliana explants by Agrobacterium tumefaciens. Plant Mol Biol 8: 291–298

Swinburne J, Balcells L, Scofield SR, Jones JDG, Coupland G (1992) Elevated levels of activator transposase mRNA are associated with high frequencies of dissociation excision in Arabidopsis. Plant Cell 4: 583–595

Tsay YF, Frank MJ, Page T, Dean C, Crawford NM (1993) Identification of a mobile endogenous transposon in Arabidopsis thaliana. Science 260: 342–344

Valvekens D, Van Montagu M, Van lijsebettens M (1988) Agrobacterium tumefaciens-mediated transformation of Arabidopsis thaliana root explants by using kanamycin selection. Proc Natl Acad Sci USA 85: 5536–5540

Van de Krol AR, Mol JNM, Stuitje AR (1988) Antisense genes in plants: an overview. Gene 72: 45–50

Van Sluys MA, Tempé J, Fedoroff N (1987) Studies on the introduction and mobility of the maize activator element in Arabidopsis thaliana and Daucus carota. EMBO J 6: 3881–3889

Vongs A, Kakutani T, Martienssen RA, Richards EJ (1993) Arabidopsis thaliana DNA methylation mutants. Science 260: 1926–1928

Voytas DF, Ausubel FM (1988) A copia-like transposable element family in Arabidopsis thaliana. Nature 336: 242–244

Zhang H, Somerville CR (1987) Transfer of the maize transposable element Mu1 into Arabidopsis thaliana. Plant Sci 48: 165–173

Zhang H, Scheirer DC, Fowle WH, Goodman HM (1992) Expression of antisense or sense RNA of an ankyrin repeat-containing gene blocks chloroplast differentiation in Arabidopsis. Plant Cell 4: 1575–1588

Developmental Regulation of Co-suppression In *Petunia hybrida*

R.B. Flavell[1], M. O'Dell[1], M. Metzlaff[1], S. Bonhomme[1,2], and P.D. Cluster[1]

1 Introduction

The gene silencing phenomenon to be discussed here, initially termed "co-suppression" (Napoli et al. 1990; Van der Krol et al. 1990; Jorgensen 1990), was observed in purple-flowered petunia plants genetically modified by the introduction of DNA containing a chalcone synthase coding sequence under the control of the strong CaMV 35S promoter and the 3' end from the nopaline synthase gene of *Agrobacterium*. The selectable marker gene consisting of the coding sequence for neomycin phosphotransferase under the control of nopaline synthase promoter and with the 3' end from the octopine synthase gene was also inserted on the same T-DNA. These genes were introduced into petunia cells via the transferred portion of the Ti plasmid of *Agrobacterium tumefaciens* (i.e. the T-DNA).

Chalcone synthase is a key enzyme in flavonoid biosynthesis and, therefore, in pigment production. These pigments are synthesized intensely in the epidermis of flower petals, but also to lesser extents in many other parts of the plant including the anthers. Pigment production is cell-type specific. Chalcone synthase gene expression is transcriptionally regulated but separate post-transcriptional effects have been described that influence the pigmentation pattern in flowers (Mol et al. 1983). In petunia, chalcone synthases are encoded by a gene family (Koes et al. 1989), and the cDNA used to create the new transgene was

[1] John Innes Centre, Norwich Research Park, Colney, Norwich, NR4 7UH, UK

[2] Present Address: INRA Centre de Versailles, Station de Genetique et Amelioration des Plantes, 78026 Versailles Cedex, France

from the chalcone synthase A (CHS A) allele responsible for most of the chalcone synthase activity in petals (Koes et al. 1989). Surprising phenotypes were produced in that a very high proportion of the first family of primary transformants had flowers with white sectors, and the flowers of some plants were completely white (Napoli et al. 1990). This was interpreted to imply that the introduction of a new chalcone synthase gene had caused the loss of most or all chalcone synthase activity from the inserted transgene and the endogenous chalcone synthase genes in the white petal sectors. This interpretation was confirmed by the correlation between the lack of anthocyanin pigment and the very low levels of mRNA from both transgene and endogenous CHS A in petals (Napoli et al. 1990; Van der Krol et al. 1990). This suppression of both kinds of homologous gene was the reason for using the term co-suppression to describe the phenomenon (Napoli et al. 1990). Subsequent analyses of large numbers of transformants and their progeny from the selfing and backcrossing of selected transgenic lines have revealed numerous important features about the co-suppression phenomenon (Jorgensen 1993b, 1994, and unpublished results).

2 Co-suppression and Plant Development

The flower phenotypes showing co-suppression have been classified on the basis of the position and extent of pigmentation in the flowers (Jorgensen 1993a,b). This classification is meaningful because phenotypes are characteristic for particular transformants even though new variants may arise, as described below. Some of the phenotypes are shown in Fig 1. They range from completely white where pigment production is suppressed in all parts of the flower—tube, corolla and anthers—to other patterns where the white segments are small. In one pattern the pigmentless sector is confined to the tube and the anthers, but frequently extends just outside the tube and to a greater extent on the lower petals. In others pigment loss occurs in small sectors along the veins and/or petal tips. In another pattern, pigment loss is orientated along the edges of the petals. The areas without pigment can be much larger in some phenotypes (Napoli et al. 1990). In yet another series of plants, the white sectors are small and dispersed across the flower in complex patterns. All these patterns point to inherent features of flower development that are revealed by the transgene. The cells which lack pigment are not simply clonally related. Instead it appears that cells occupying certain locations in the floral meristem with respect to architectural features of petal shape such as lines of symmetry, respond similarly (but not identically) from petal to petal to the presence of the transgene, and these responses are different from those of other cells in other positions. The pattern boundaries are coincident in the upper and lower epidermis. Thus pattern formation may also require intercellular communication.

The untransformed parent plants show no evidence of such pigmentation patterns, though other varieties do (Red Star and Velvet Picotee; Mol et al. 1983;

Fig. 1. Flower patterns resulting from insertion of the transgene consisting of the coding sequence of CHS A under the control of the CaMV 35S promoter (see colour version of the figure on the backcover). The untransformed parent has only purple flowers. Phenotypes from top middle round to bottom left display: extensive co-suppression emanating from lower petal junctions, complete co-suppression, co-suppression from lower petal junctions, co-suppression along petal veins

VAN BLOKLAND 1994). From the principles of flower design, one can assume that the architectural basis of the patterns is not caused by the transgene but is an inherent feature of flower development. However, elements of this feature somehow interact with the transgene or its product to produce the observed patterns, and different versions (states) of the transgene interact differently to create the different patterns (JORGENSEN 1993a,b).

The observation that many transgenic plants display a characteristically patterned flower phenotype, based on the patterns of co-suppression, implies that the 'state' of the transgene is somatically inherited. When the meiotic inheritance of transgene effects on flower phenotypes was examined (JORGENSEN 1993b and unpublished), several outcomes were noted. In many cases, the phenotype bred true and is thus germinally stable. In other plants examined, a new range of somatically inherited phenotypes was observed. For example, from a backcross between a white transformant containing two tandem copies of the new genes and its untransformed parent, many phenotypes were obtained including fully purple, fully white and various patterned types. In these cases the phenotype based on the floral positions of cells showing co-suppression is germinally unstable and the transgene presumably alters its state. Thus it can be concluded that a given transgene can exist in different epiallelic states, and these states can change during meiosis or early embryonic development (JORGENSEN 1993b). Occasionally a lateral branch emerges that displays a different flower phenotype with more or less pigmentation, and the variation is inherited, implying that a change has occurred in the L2 layer of cells in the flowers (JORGENSEN 1994 and personal communication).

Petunias produce flowering branches from organised groups of cells (meristems) in the axils of leaves or on the flanks of meristems. A genetically different branch results if the group of cells in the meristem flank becomes modified. Occasionally single variant flowers, gradients of phenotypic change as a branch ages, and simultaneous changes in different branches have been noted (JORGENSEN 1994) implying that changes can occur in any floral meristem. Because the inherited L2 layer and the L1 layer in which epidermal pigment is produced are separate developmental lineages of cells, it is reasonable to consider the possibility that the changes in transgene state behind pattern changes occur in many cells of a meristem essentially simultaneously.

The remainder of this chapter deals with the origins of the pigmentless phenotype created by the insertion of the CHS A coding sequence under the control of the CaMV 35S promoter.

3 Hypotheses to Explain Gene Silencing

Numerous examples are known, in at least six plant species, where gene inactivations results from the introduction of additional homologous sequences. These have been reviewed elsewhere (JORGENSEN 1990, 1991, 1992; MOL et al. 1991; KOOTER and MOL 1993; MATZKE and MATZKE 1993; MATZKE et al. 1993; ASSAAD et al. 1993; VAUCHERET 1993; GORING et al. 1991; MEYER et al. 1993; GRIERSON et al. 1991; FLAVELL 1994; HART et al. 1992; MEINS 1989; MEINS and KUNZ 1994), and in other chapters in this book (for example see Hamilton et al. and de Lange et al., this volume). They will not therefore be discussed extensively here. However, it should be noted that no single mechanism can explain the variety of examples where loss of gene expression has occurred.

Four kinds of hypotheses have been put forward to explain the diversity of gene silencing phenomena. In the first, inactivation of transcription is postulated due to the physical interaction (ectopic pairing) in the nucleus of the duplicated but non-allelic sequences (loci). Cycles of DNA–DNA or chromatin–chromatin interactions (see Fig. 2) could leave the chromatin structure or methylation patterns of the participating genes in different states which could consequently interfere with

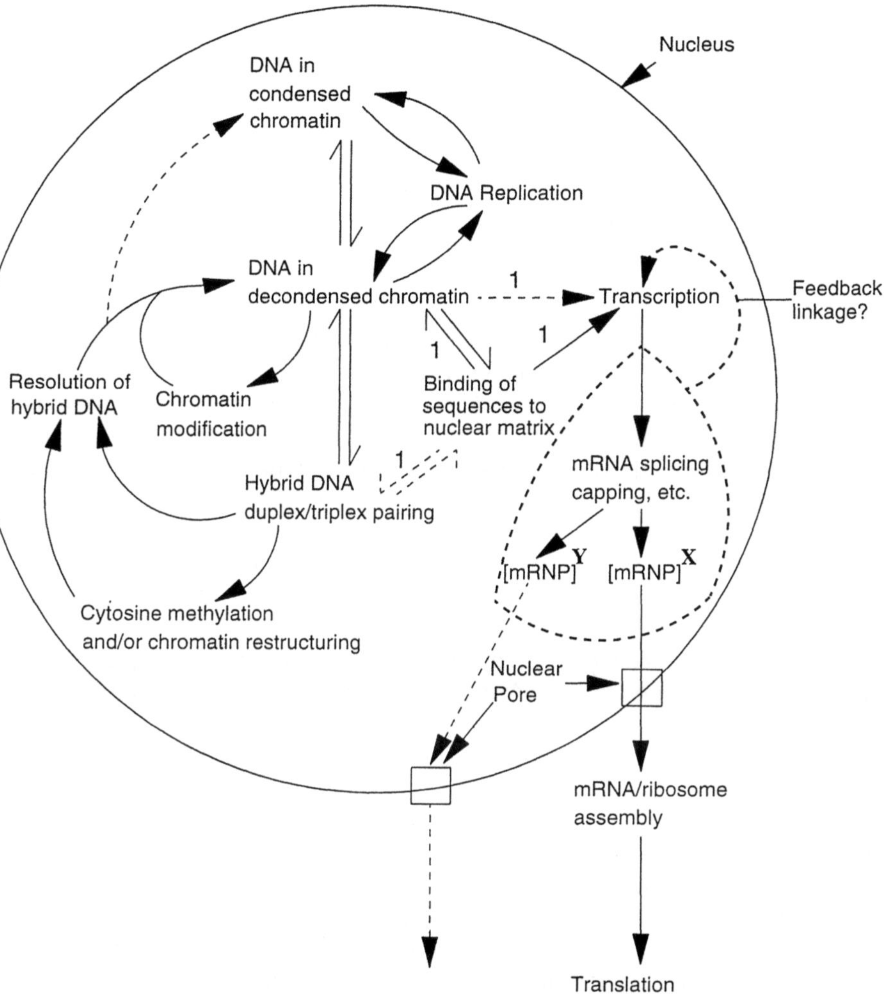

Fig. 2. Cellular processes relevant to models for gene silencing. The network of pre-transcriptional events illustrates how the structure of decondensed chromatin, the substrate for transcription, can be modified by various sorts of changes including cytosine methylation and interactions with homologues including transgenes. The changes could modify decondensed chromatin such that it does not bind properly to the nuclear matrix or bind transcription complexes efficiently. After "normal" transcription mRNA is processed, capped and polydenylated in messenger RNA nuclear protein particles $[mRNP]^X$ which are then exported from the nucleus and the mRNA traslated on ribosomes. Where gene silencing is post-transcriptional, transgene mRNA processing, splicing, capping or polyadenylation could be aberrant, thereby leading to synthesis of aberrant particles $[mRNP]^Y$. These might not be translated efficiently and may be substrates for RNases. They may also be substrates for antisense RNA formation. In any event they do not give rise to protein product. There is the possibility of aberrant mRNA production influencing transcription. Further details are described in the text

the assembly of essential transcription complexes or the binding of the chromatin to the nuclear matrix. These processes are labelled 1 in Fig. 2. There is no direct evidence for such interactions occurring in plants, but precedents come from studies on fungi and *Drosophila*. In *Neurospora* and *Ascobolus*, DNA homology-searching processes and hybrid DNA formation have been inferred from the inactivation of duplicated sequences via cytosine methylation in premeiotic cells (Selker 1990; Foss and Selker 1991; Rhounim et al. 1992; Faugeron et al. 1990). In yeast, equivalent frequencies of allelic and ectopic meiotic recombination have also been taken to imply the existence of efficient, generalised, DNA sequence homology searching processes (Haber et al. 1991). In *Drosophila* there are many examples where expression of a gene is influenced by "sensing" the presence of another specific gene after some kind of localised somatic chromosome pairing. The pairing could be mediated via DNA, RNA or transcription complexes. The consequentially altered chromatin, sometimes heterochromatic, state created following the interlocus interactions can be clonally inherited when not disturbed by other events (Tartof and Henikoff 1991; Henikoff 1992; Pirotta 1990; Wu 1993; Paro 1990).

The second hypothesis is based upon elevated competition between the increased number of genes for non-diffusible sequence-specific factors essential for ordered transcription or translation.

The third hypothesis focuses on post-transcriptional events. It postulates the degradation of the specific mRNAs due to the synthesis of homologous antisense RNAs in the cell, formation of double-stranded RNAs between the antisense RNA and mRNAs and recognition of the aberrant duplexes as substrates for a RNase. Mutual inactivation of homologous mRNAs can often be achieved by the introduction of antisense gene. Double-stranded RNAs may also inhibit translation if they are formed in the cytoplasm (Temple et al. 1993; Cornelissen and Vandewiele 1989). Evidence for the existence of dsRNA in plants is however very weak (Grierson et al. 1991; Jorgensen 1991; Mol et al. 1991). These antisense RNAs could be made from an unknown promoter close to the transgene functioning in the appropriate orientation, possibly by readthrough from a neighbouring gene or by the action of RNA-dependent RNA polymerase on aberrantly accumulated mRNAs (Lindbo et al. 1993; Flavell 1994). This latter enzyme exists in plant cells.

The fourth hypothesis postulates the inhibition of transcription and/or translation by feedback from a specific gene product that accumulates in aberrantly high concentrations in the transgenic plants. This would constitute a self-induced, autoregulated control system (Hart et al. 1992; Meins 1989; Meins and Kunz 1994).

4 Mechanisms and Hypotheses for Co-suppression of Chalcone Synthase in Transgenic Petunias

We now consider the petunia chalcone synthase case in the light of these hypotheses. Throughout this discussion, it is relevant to bear in mind that not all transgenic plants containing CHS A transgenes display co-suppression in the petals, and that in some plants only specific segments of petals show co-suppression. Also, it is important to remember that inherited somatic and meiotic changes can occur to influence the extent to which co-suppression is observed.

In the process of making the transgenic plants, it can be expected that different numbers of T-DNAs become stably inserted into different petunia plants and display different structures. Tandem arrays of T-DNAs are common, as are copies inverted with respect to one another. Genetically unlinked T-DNAs also accumulate. Thus plants with different numbers of active genes are likely to be produced, as noted in other studies on transgenic plants (Hobbs et al. 1993; Assaad et al. 1993; Scheid et al. 1991; Linn et al. 1990). It will be important to investigate thoroughly whether the structure of T-DNA inserts influences the extent of co-suppression and the kinds of flower pattern produced. Van Blokland (1994) has concluded that phenotypic effects of CHS A transgenes are correlated with the presence of inverted repeats of T-DNA.

Where there are multiple copies of the chalcone synthase transgene then the copies might interact (see Fig. 2) to silence transcription of one another and the endogenous CHS A genes. Such silencing has been recorded for several sorts of transgenes (Peach and Velten 1991; Assaad et al. 1993; Hobbs et al. 1990, 1993; Elkind et al. 1990; Linn et al. 1990; Matzke et al. 1994b; Vaucheret 1993). Inverted repeats seem to be more frequently associated with transcriptional silencing (Hobbs et al. 1993). How such physical interactions occur is unknown, but they may be the means whereby one or more of the duplicated sequences gain some methylated cytosines. No evidence for silenced CHS A genes becoming routinely methylated has yet been obtained in investigations of several sites within the coding sequences and promoters.

There is evidence, however, that in some transgenic petunias, CHS A transcription is not blocked in petal cells showing loss of pigment. Run-on transcription assays on nuclei from isolated purple and white petal sectors from the same plant show similar levels of CHS A transgene and endogenous CHS A transcription (Van Blokland 1994). Furthermore, similar levels of unprocessed nuclear endogenous CHS A transcripts have been detected in flowers of some co-suppressed and non co-suppressed variant plants in our laboratory and in that of Mol and co-workers. The levels of RNA transcribed in isolated petal nuclei are not correlated with the extent of chalcone synthase suppression (Van Blokland 1994; Kooter and Mol 1993; Mol et al. 1991). These details are reviewed in another chapter in this book (de Lange et al., this volume). We have also found in some plants that white flower sectors retain high levels of CHS A RNA, making it likely that post-transcriptional losses of functional mRNAs are the cause of or a major contributor to the co-suppression phenotype. Studies of inactivation of some other transgenes in plants have also concluded that the inactivation is post-transcriptional (SMITH et al. 1990a; DE CARVALHO et al. 1992; BATE et al. 1992; MEINS and KUNZ 1994).

In plants where transcription of the CHS A transgenes is not blocked but steady state functional mRNA levels are very low, then a major cause of co-suppression could be accumulation of excess levels of antisense RNAs to the CHS A mRNA, double-stranded RNA formation and degradation of the duplex RNA (third hypothesis above). The presence of antisense RNA to chalcone synthase has been investigated in transgenic petal tissues differing in co-suppression, i.e. purple and white. Of the particular variants studied by us, most were derived from the same transgenic parent and possess two copies of the transgene in inverted orientation. Reverse transcriptase and primers specific for antisense RNA were used to make DNA copies of RNA in RNA extracts isolated from white or purple sectors. Antisense chalcone synthase RNAs were found in both white and purple flower sectors but only in transgenic plants. It is, therefore, concluded that the antisense RNAs are due to the transgene. The finding that antisense chalcone synthase RNAs are in both white and purple sectors suggest that if antisense RNA is essential for the loss of mRNA and gene expression in this genotype, it is clearly insufficient. Similar conclusions have been drawn by Mol and co-workers (de Lange et al, this volume; VAN BLOKLAND 1994) who used other assays to detect antisense RNA.

How is antisense RNA produced from the transgene, what is its structure, and how does it function? These important questions still have to be examined experimentally. It will be necessary to examine many different transgenic plants with different numbers and kinds of transgene structures since it is not clear how antisense RNA could be produced so efficiently in all transformants (JORGENSEN 1991). Where antisense RNA is not transcribed from defined genomic promoters it could be formed by an RNA-dependent RNA polymerase using sense mRNA as template (LINDBO et al. 1993; FLAVELL 1994).

On simple considerations of how antisense RNA interferes with sense mRNA, it would be assumed that the higher the antisense to sense RNA ratio, the more efficient would be the loss of sense gene expression. Some data in plants to support this have been produced (SMITH et al. 1990b; HAMILTON et al. 1990; CANNON et al. 1990; ROBERT et al. 1989; VAN DER MEER et al. 1992). However, there are many reports of discrepancies between the relative levels of antisense RNA transcripts and loss of sense gene expression (reviewed in de Lange, this volume; CANNON et al. 1990; STOCKHAUS et al. 1990; VAN DER KROL et al. 1988). VAN BLOKLAND (1994) found in petunias transgenic for chalcone synthase that antisense transcription could be high in the absence of co-suppression or vice versa. If antisense RNA is the cause of degradation of CHS A mRNAs, but overall steady state or transcription levels of antisense mRNA do not correlate with co-suppression it must be a small fraction of the antisense RNA that is critical, and this fraction must have efficient access to the unprocessed primary RNA transcripts or mRNAs formed after processing, capping and poly-A tail additions. This implies that the variation in co-suppression in transgenic plants that makes antisense and sense RNAs could involve variation in the accessibility of antisense and sense RNAs to each other in the nucleus or the cytoplasm (see later).

Post-transcriptional loss of CHS A gene expression and pigment production could, alternatively, be due to the accumulation of excess levels of CHS mRNA and the consequential induction of an mRNA-specific process, that is able to catalyse the inactivation and/or degradation of transgene and endogenous CHS A mRNAs (fourth hypothesis above). This hypothesis includes the notion of critical localised threshold levels of mRNA in a cell. Where mRNA levels are below the threshold, purple pigment is produced; in contrast, when the level is exceeded, active mRNA is lost and no pigment is produced. There are several predictions from this model.

1. In plants where all flower epidermal cells lack pigment, the threshold mRNA concentrations are likely to have been exceeded in all cells of the plant. This is as observed in many white flowered plants where levels of transgene mRNA are very low in stems and leaves, as well as flowers (NAPOLI et al. 1990; this laboratory, unpublished).
2. In plants which have purple flowers with white sectors, the levels of localised active mRNA is likely to be higher than that found in plants that make only purple flowers. This is because in such plants small increases in the level of mRNA accumulated would exceed the threshold more readily and thus lead to the pigment loss. Some evidence has been gained to support this hypothesis in that the levels of transgene CHS mRNA accumulating in leaves of transgenic plants with purple and white flowers is greater than in leaves of transgenic plants forming only purple flowers (unpublished results).

If localised mRNA concentrations are the determinant of the post-transcriptional trigger for co-suppression, then the critical parameters affecting co-suppression would be the rate of transgene transcription and/or changes in the efficiency (rate) of mRNA transport through the nucleus, of export through the nuclear envelope, of binding to the ribosome and of translation. Changes in flower pigment production due to different levels of translatable CHS A mRNA could therefore come about through (a) pre-transcriptional events including changes in the levels of transcription factors, restructuring of chromatin, (and/or) changed cytosine methylation to affect the affinity of the transgenes for transcription complexes as noted earlier (see Fig. 2), and/or (b) changes in the rate of mRNA transport, etc. The latter could result from the transgene altering its position in the nucleus with respect to nuclear transport channels and the supply of protein components of the mRNPs essential for correct mRNA processing, transport and export (FLAVELL 1994). Furthermore, aberrantly high levels of mRNA in the nucleus might lead to mRNP particles with a different complement of proteins (WOLFFE 1994) from those formed when the CHS genes are optimally transcribed for mRNA processing and transport and when nuclear mRNA levels are much lower (see Fig. 2). Such modified mRNPs might not make the mRNAs available for translation.

A growing number of proteins are known that bind to mRNA and prevent translation as part of specific regulatory mechanisms. The studies on the ubiquitous Y box proteins and the FRG Y proteins of Xenopus oocytes, in particular, are interesting (WOLFFE 1994). These proteins bind to mRNAs and inhibit their translation (BOUVET and WOLFFE 1994). They will bind to a range of double and single stranded DNAs and RNAs including specific Y box sequences in gene promoters (reviewed in WOLFFE 1994). They also stimulate mRNA synthesis, but not necessarily by binding to the promoter motifs. FRG Y2 (a predominantly nuclear Y box protein) has been immunolocalised to nascent transcripts on lampbrush chromosomes (SOMMERVILLE et al. 1993). These observations are consistent with a direct role for FRG Y2 in packaging mRNA in the nucleus and for somehow linking the competence of the mRNA for translation with its transcription (BOUVET and WOLFFE 1994). Recruitment of some other heterogeneous nuclear (hn) RNP proteins on to pre-mRNA is also dependent on transcription (PINAL-ROMA and DREYFUSS 1992), and there is a growing list of eukaryotic proteins having dual roles in the transcription and translation processes (reviewed in BOUVET and WOLFFE 1994).

If protein-based regulatory systems linking transcription, mRNA packaging and mRNA translatability are present in plants, then they might be responsible for post-transcriptional loss of gene expression following "aberrant" mRNA synthesis from the CHS A transgenes. It is also possible that they could lead to shut down of transcription. This whole area of nuclear biology needs to be explored in the context of understanding how aberrant active CHS A transgenes can promote loss of pigment production in flower petals. While it is possible to imagine how mRNA from an aberrantly located transgene might be sequestered into an inactive mRNP structure, how would this affect mRNAs from the endogenous CHS A genes? Perhaps the protein-mediated regulatory systems could also provoke cross-talk between homologous mRNP complexes and sequester all CHS A RNAs into aberrant mRNPs.

If excess CHS mRNAs accumulate due to higher rates of transcription or low rates of mRNP maturation in the nucleus and/or translation, how are they degraded? This could result from the aberrant mRNPs being recognised by RNases and the RNAs consequently degraded (SACHS 1993; SULLIVAN and GREEN 1993). Alternatively, antisense RNA could be produced on the accumulated mRNA templates by RNA-dependent RNA polymerase, and these double-stranded RNA structures are subsequently degraded. Cycles of RNA production of both antisense and sense RNAs could emerge from this process to provide an autocatalytic system for the production ds RNAs. Any such antisense RNAs could, of course, explain loss of both transgene and endogenous CHS A mRNAs.

5 Concluding Perspective

From surveying the range of examples of gene silencing, it is clear that multiple mechanisms contribute to the observed phenotypes and in some examples mechanisms resulting in inhibition of transcription are major determinants, while in others post-transcriptional events occur. This diversity of mechanism may also appear between plants genetically altered by insertion of the same or related transgenes, but at different sites, in different arrangements and with different effective promoter strengths. Thus it is desirable not to automatically lump all petunias involving CHS A transgenes into a homogeneous group and attempt to find a single mechanism for the observed gene silencing or lack of it.

The scenario described above for the post-transcriptional control of chalcone synthase silencing may also be combined with, or lead to, variable patterns of transcription silencing in different genotypes. The data accumulated to date point to association of the phenomena with higher levels of mRNA synthesis and/or antisense RNA, and it has been argued that there are probably more than one cellular pool or RNP package for each of these molecules. The fact that sense mRNAs, may exist in different RNP pools and packages implies that we need to look for the different structural forms that may have different stabilities and opportunities to associate with ribosomes and be translated. Similarly we need to investigate whether different antisense RNPs exist.

If only one of the antisense RNP pools is available to interact with only one of the classes of sense mRNP, then the interacting classes are likely to be degraded in co-suppressed tissues, while other classes or pools might not be. Such discoveries might help explain the lack of correlation between antisense RNA to sense mRNA ratios and co-suppression phenotypes in different plants and tissues.

Hypotheses that propose the formation of different mRNP packages from active transgenes in aberrant nuclear positions and the production of pools of antisense RNA in some cases offer the following sorts of explanations for the origins of purple and white flower sectors: In transgenic plants where CHS A transgene transcripts are efficiently processed, packaged and exported then aberrant mRNPs would not accumulate and so co-suppression would not occur. Such plants would have purple flowers. If antisense RNPs were produced in such plants, the antisense RNA might not be accessible to the sense mRNPs and so the flowers would be purple. If, however, transgene mRNA were processed, packaged and exported inefficiently, due to the location or other features of transgene chromatin, then critical levels of nuclear mRNA would be exceeded, packaging could be aberrant and a different mRNP structure for all CHS A mRNAs might result. Messenger RNA in this structure might not be translated, or might be accessible to RNases and antisense RNAs or to RNA-dependent RNA polymerase that makes antisense RNA. Any of these would result in the formation of white flowers.

These ideas are testable and imply that switches in pigment production during transgenic CHS A plant development could result from (a) a change in nuclear position of the transgene; (b) a change in transcription rates (these would constitute inherited changes in state of a transgene); (c) a change in cell physiology influencing nuclear processing, RNA packaging, export and mRNA translation rates; and/or (d) a change in antisense RNA synthesis. Variation in such parameters would not be surprising during meristem development, growth in different environments or during specific developmental phases. The patterns in flowers are presumably due to similar changes.

The developmental changes influencing the nuclear metabolism of specific clusters of cells in floral meristems and floral tissues are unknown. However, analyses of the CaMV 35S promoter have revealed that it contains multiple elements that respond differently in different petunia floral tissues, leading to differential transcription (Benfey and Chua 1989; Benfey et al. 1989). Thus the enhanced probability that co-suppression occurs in the flower tube and veins in some transgenic genotypes containing CHS A under the control of the CaMV 35S promoter or in other regions in other genotypes could be due to differential interactions between the promoter and enhancer structures of the promoter and the transcription factor concentrations present in different sectors of the meristem and floral tissues.

In conclusion, it is clear that studies into the origins of co-suppression and gene silencing in general will teach us many new features of cell biology and the control of gene expression. Furthermore, because of the wish to create agriculturally novel transgenic plants, understanding how active transgenes can lead to gene silencing is of considerable commercial interest.

Acknowledgements. We are grateful to John Bedbrook of DNAP, Oakland, California, USA, for the petunia transformants used in this study and especially to Rich Jorgensen, University of California, Davis, USA, for extensive collaborations and discussions in this project area.

References

Assaad FF, Tucker KL, Signer ER (1993) Epigenetic repeat-induced gene silencing (RIGS) in Arabidopsis. Plant Mol Biol 22: 1067–1085

Bate NJ, Pallas JA, Elkind Y, Lamb CJ (1992) Sense-suppression of endogenous phenylalanine ammonia-lyase (PAL) expression in tobacco containing a bean PAL transgene. Abstract 9, The Tenth John Innes Symposium, The Chromosome, Norwich, September 1992

Benfey PN, Chua N-H (1989) Regulated genes in transgenic plants. Science 244: 174–181

Benfey PN, Ren L, Chua N-H (1989) The CaMV 35S enhancer contains at least two domains which can confer different developmental and tissue-specific expression patterns. EMBO J 8: 2195–2202

Bouvet P, Wolffe AP (1994) A role for transcription and FRGY2 in masking maternal mRNA within Xenopus oocytes. Cell 77: 931–941

Cannon M, Platz J, O'Leary M, Sookdeo C, Cannon F (1990) Organ-specific modulation of gene expression in transgenic plants using antisense RNA. Plant Mol Biol 15: 39–47

Cornelissen M, Vandewiele M (1989) Both RNA level and translation efficiency are reduced by antisense RNA in transgenic tobacco. Nucl Acids Res 17: 833–843

de Carvalho F, Gheysen G, Kushnir S, van Montagu M, Inze D, Castresana C (1992) Suppression of beta-1,3-glucanase transgene expression in homozygous plants. EMBO J 11: 2595–2602

Elkind Y, Edwards R, Mavandad M, Hedrick SA, Ribak O, Dixon RA, Lamb CJ (1990) Abnormal plant development and down-regulation of phenylpropanoid biosynthesis in transgenic tobacco containing a heterologous phenylalanine ammonia-lyase gene. Proc Natl Acad Sci USA 87: 9057–9061

Faugeron G, Rhounim L, Rossignol J-L (1990) How does the cell count the number of ectopic copies of a gene in the premeiotic inactivation process acting in Ascobolus immersus? Genetics 124: 585–591

Flavell RB (1994) Inactivation of gene expression in plants as a consequence of novel sequence duplication. Proc Natl Acad Sci USA 91: 3490–3496

Foss HM, Selker EU (1991) Efficient DNA pairing in a Neurospora mutant defective in chromosome pairing. Mol Gen Genet 231: 49–52

Goring DR, Thomson L, Rothstein SJ (1991) Transformation of a partial nopaline synthase gene into tobacco suppresses the expression of a resident wild-type gene. Proc Natl Acad Sci USA 88: 1770–1774

Grierson D, Fray RG, Hamilton AJ, Smith CJS, Watson CF (1991) Does co-suppression of sense genes in transgenic plants involve antisense RNA. Trends Biotech 9: 122–123

Haber JE, Leung W-Y, Borts RH, Lichten M (1991) The frequency of meiotic recombination in yeast is independent of the number and position of hologous donor sequences: implications for chromosome pairing. Proc Natl Acad Sci USA 88: 1120–1124

Hamilton AJ, Lycett GW, Grierson D (1990) Antisense gene that inhibits synthesis of the hormone ethylene in transgenic plants. Nature 346: 284–287

Hart CM, Fischer B, Neuhaus J-M, Meins Jr. F (1992) Regulated inactivation of homologous gene expression in transgenic Nicotiana sylvestris plants containing a defense-related tobacco chitinase gene. Mol Gen Genet 235: 179–188

Henikoff S (1992) Position effect and related phenomena. Current Op-Genet Develop 2: 907–912

Hobbs SLA, Kpodar P, DeLong CMO (1990) The effect of T-DNA copy number, position and methylation on reporter gene expression in tobacco transformants. Plant Mol Biol 15: 851–864

Hobbs SLA, Warkentin TD, DeLong CMO (1993) Transgene copy number can be positively or negatively associated with transgene expression. Plant Mol Biol 21: 17–26

Jorgensen R (1990) Altered gene expression in plants due to trans interactions between homologous genes. Trends Biotech 8: 340–344

Jorgensen R (1991) Beyond antisense—how do transgenes interact with homologous plant genes? Trends Biotechnol 9: 266–267

Jorgensen R (1992) Silencing of plant genes by homologous transgenes. Agbiotech News Info 4: 265N–273N

Jorgensen R (1993a) The germinal inheritance of epigenetic information in plants. Phil Trans R Soc Lond B 339: 173–181

Jorgensen R (1993b) Elicitation of organised pigmentation patterns by a chalcone synthase transgene. In: Amasino RM (ed) Cellular communication in plants. Plenum, New York, pp 87–92

Jorgensen R (1994) The significance of epigenetically impositions on the plant genome: a paragenetic function for chromosomes. Dev Genet (in press)

Koes RE, Spelt CE, Elzen PJMvd, Mol JNM (1989) Cloning and molecular characterization of the chalcone synthase multigene family of Petunia hybrida. Gene 81: 245–257

Kooter JM, Mol JNM (1993) Trans-inactivation of gene expression in plants. Curr Op Biotechnol 4: 166–171

Lindbo JA, Silva-Rosales L, Proebsting WM, Dougherty WG (1993) Induction of a highly specific antiviral state in transgenic plants: implications for gene regulation and virus resistance. Plant Cell 5: 1749–1759

Linn F, Heidmann L, Saedler H, Meyer P (1990) Epigenetic changes in the expression of the maize A1 gene in Petunia hybrida: role of numbers of integrated gene copies and state of methylation. Mol Gen Genet 222: 329–336

Matzke MA, Matzke AJM (1993) Genomic imprinting in plants: parental effects and trans-inactivation phenomena. Annu Rev Plant Physiol Plant Mol Biol 44: 53–76

Matzke MA, Neuhuber F, Matzke AJM (1993) A variety of epistatic interactions can occur between partially homologous transfer loci brought together by sexual crossing. Mol Gen Genet 236: 379–386

Matzke MA, Matzke AJM, Scheid OM (1994a) Inactivation of repeated genes—DNA–DNA interaction? In: Paszkowski J (ed) Homologous recombination in plants. Kluwer, Amsterdam, pp 271–307.

Matzke AJM, Neuhuber F, Park Y-D, Ambors PF, Matzke MA (1994b) Homology-dependent gene silencing in transgenic plants: epistatic silencing loci contain multiple copies of methylated transgenes. Mol Gen Genet 244: 219–229

Meins F Jr. (1989) Habituation: heritable variation in the requirement of cultured plant cells for hormones. Annu Rev Genet 23: 395–408

Meins Jr. F, Kunz C (1994) Silencing of chitinase expression in transgenic plants: an autoregulatory model. In: Paszkowski J (ed) Gene inactivation and homologous recombination in plants. Kluwer, Amsterdam (in press).

Meyer P, Heidmann I, Niedenhof I (1993) Difference in DNA methylation are associated with a paramutation phenomenon in transgenic petunia Plant J 4: 89–100

Mol JNM, Schram AW, de Vlaming P, Gerats AGM, Kreuzaler F, Hahlbrock K, Reif HJ, Veltkamp E (1983) Regulation of flavonoid gene expression in Petunia hybrida: description and partial characterization of conditional mutant in chalcone synthase gene expression. Mol Gen Genet 192: 424–429

Mol J, Van Blokland R, Kooter J (1991) More about co-suppression. Trends Biotech 9: 182–183

Napoli C, Lemieux C, Jorgensen R (1990) Introduction of a chimeric chalcone synthase gene into petunia results in reversible co-suppression of homologous genes in trans. Plant Cell 2: 279–289

Paro R (1990) Imprinting a determined state into the chromatin of Drosophila. Trends Genet 6: 416–421

Peach C, Velten J (1991) Transgene expression variability (position effect) of CAT and GUS reporter genes driven by linked divergent T-DNA promoters. Plant Mol Biol 17: 49–60

Pinol-Roma S, Dreyfuss G (1992) Shuttling of pre-mRNA binding proteins between nucleus and cytoplasm. Nature 355: 730–732

Pirotta V (1990) Transvection and long distance regulation. BioEssays 12: 409–414

Rhounim L, Rossignol J-L, Faugeron G (1992) Epimutation of repeated genes in Ascobolus immersus. EMBO J 11: 4451–4457

Robert LS, Donaldson PA, Ladaique C, Altosaar I, Arnison PG, Fabijanski SF (1989) Antisense RNA inhibitors of beta-glucuronidase gene expression in transgenic tobacco plants. Plant Mol Biol 13: 399–409

Sachs AB (1993) Messenger RNA degradation in eukaryotes. Cell 74: 413–421

Scheid OM, Paszkowski J, Rothstein SJ (1991) Reversible inactivation of transgene in Arabidopsis thaliana. Mol Gen Genet 228: 104–112

Selker EU (1990) Premeiotic instability of repeated sequences in Neurospora crassa. Ann Rev Genet 24: 579–613

Smith CJS, Watson CF, Bird CR, Ray J, Schuch W, Grierson D (1990a) Expression of a truncated tomato polygalacturonase gene inhibits expression of the endogenous in transgenic plants. Mol Gen Genet 224: 477–481

Smith C, Watson CF, Morris PC, Bird CR, Seymour GB, Gray JE, Arnold C, Tucker GA, Schuch W, Harding S, Grierson D (1990b) Inheritance and effect on ripening of antisense polygalacturonase genes in transenic tomatoes. Plant Mol Biol 14, 369–379

Sommerville J, Baird J, Turner BM (1993) Histone A4 acetylation and transcription in amphibian chromatin. J Cell Biol 120: 277–290

Stockhaus J, Hofer R, Renger G, Westhoff P, Wydrzynski T, Willmitzer L (1990) Anti-sense RNA efficiently inhibits formation of the 10Kd polypeptide of photosystem-II in transgenic potato plants - analysis of the role of the 10Kd protein. EMBO J 9: 3013–3021

Sullivan ML, Green PJ (1993) Post-transcriptional regulation of nuclear-encoded genes in higher plants: the roles of mRNA stability and translation. Plant Mol Biol 23: 1091–1104

Tartof KD, Henikoff S (1991) Trans-sensing effects from Drosophila to humans. Cell 65: 201–203

Temple SJ, Knight TJ, Unkefer PJ, Senguptagopalan C (1993) Modulation of glutamine synthetase gene expression in tobacco by the introduction of an alfalfa glutamine synthetase gene in sense and antisense orientation—molecular and biochemical analysis. Mol Gen Genet 236: 316–325

Van Blokland R (1994) Trans-inactivation of flower pigmentation genes in Petunia hybrida. Ph.D thesis, Vrijie Universiteit Amsterdam

Van der Krol, AR Lenting PE, Veenstra J, Van der Meer IM, Koes RE, Gerats AGM, Mol JNM, Stuitje AR (1988) An anti-sense chalcone synthase gene in transgenic plants inhibits flower pigmentation. Nature 333: 866–869

Van der Krol AR, Mur LA, Beld M, Mol JNM, Stuitje AR (1990) Flavonoid genes in petunia addition of a limited number of gene copies may lead to a suppression of gene expression. Plant Cell 2: 291–299

Van der Meer LM, Stam M, Van Tunen AJ, Mol JNM, Stuitje AR (1992) Antisense inhibition of flavonoid biosynthesis in Petunia anthers results in male sterility. Plant Cell 4: 253–262

Vaucheret H (1993) Identification of a general silencer for 19S and 35S promoters in a transgenic tobacco plant.: 90 bp of homology in the promoter sequence are sufficient for trans-inactivation. C R Acad Sci Paris 316: 1471–1483

Wolffe AP (1994) Structural and functional properties of the evolutionary ancient Y-box family of nucleic acid binding proteins. BioEssays 16: 245-251

Wu C-T (1993) Transvection, nuclear structure and chromatin proteins. J Cell Biol 120: 587–590

Suppression of Flavonoid Flower Pigmentation Genes in *Petunia hybrida* by the Introduction of Antisense and Sense Genes

P. DE LANGE[1,2], R. VAN BLOKLAND[1], J.M. KOOTER[1] and J.N.M. MOL[1]

1 Introduction

Introduction of antisense and sense transgenes in higher plants can lead to the suppression of the homologous resident gene(s). Over the past 6 years a large number of transgenic plants containing sense and antisense transgenes have

[1] Department of Genetics, Vrije Universiteit, Institute for Molecular Biological Sciences, BioCentrum Amsterdam, De Boelelaan 1087, 1081 HV Amsterdam, The Netherlands
[2] Current address: Department of Endocrinology and Reproduction, Faculty of Medicine and Health Sciences, Erasmus University, P.O. Box 1738, 3000 DR Rotterdam, The Netherlands

been generated for various purposes. First, antisense and sense genes have been introduced into commercial crops and ornamental plant varieties to suppress the expression of resident genes with the aim of modifying or improving plant performance. Second, the function of several cryptic genes has been determined by manipulating their expression with the help of antisense genes. Finally, the use of antisense and sense gene suppression has broadened our knowledge of control steps in physiological processes (see VAN BLOKLAND et al. 1993 and MOL et al. 1994 for reviews). Despite the success of these strategies, the mode of action of antisense and sense genes is still poorly understood.

Antisense-mediated control has been observed in bacteria, fungi, plants and mammalian systems. The preferred model for antisense RNA action involves the formation of double-stranded RNA (dsRNA) with complementary RNAs. dsRNA has indeed been detected in *Escherichia coli* in the case of plasmid copy number control (CESARENI et al. 1991) and in transformed *Drosophila* cells (BUNCH and GOLDSTEIN 1989). *Xenopus* oocytes (MELTON 1985; BASS and WEINTRAUB 1987) and mammalian cells (KIM and WOLD 1985; KRYSTAL et al. 1990). The dsRNA form may prevent RNAs from being processed, transported or translated or induce rapid degradation. Attempts to detect dsRNA molecules in plants have been unsuccessful so far (R. van Blokland, unpublished results and D. Grierson, personal communication). This might indicate that such molecules are extremely unstable in plants or, alternatively, that antisense genes work via a different mechanism. Experiments discussed in this chapter indicate that the generally accepted dsRNA model is far too simple and that additional factors play an important role in antisense gene-mediated silencing.

Unexpectedly, introduction of sense transgenes in plants sometimes leads to gene silencing. The mechanism by which this occurs can vary, depending on the nature and the origin of the transgene. For instance, in some cases interaction between transgenes is associated with DNA methylation and transcriptional silencing (see the chapters by A.J.M and M.A. Matzke, P. Meyer and O. Mittelsten Scheid, this volume). In other cases where the interaction between transgenes and homologous resident genes is studied, post-transcriptional silencing is observed (see the chapter by A.J. Hamilton et al. and F. De Carvalho Niebel et al., this volume, and below). However, in all cases studied, the homology between the interacting genes is of great importance.

Gene silencing can occur without transgenes. Examples are the post-transcriptional silencing of *chs* genes in the petunia varieties Red Star and Velvet Picotee (VAN BLOKLAND 1994), the transcriptional silencing of the *B* gene in maize (see PATTERSON et al. 1993 and the chapter by G.I. Patterson and V.L. Chandler, this volume) and the suppression of *chs* in snapdragon by a rearranged *chs* allele (BOLLMANN et al. 1991).

The flavonoid biosynthesis route of *Petunia hybrida* offers a convenient system for study of the interaction between sense or antisense transgenes and homologous resident genes, as changes in flavonoid pigment gene expression are directly visible. In this chapter we will focus on the use of this system to study the mechanistic aspects of gene silencing. Several detailed and broader reviews on gene silencing have been published elsewhere (GRIERSON et al. 1990; JORGENSEN

1990, 1992; KOOTER and MOL 1993; MATZKE and MATZKE 1993; NELLEN and LICHTENSTEIN 1993; VAN BLOKLAND et al. 1993; MOL et al. 1994).

2 Gene Silencing by Antisense Genes

2.1 Examples of Gene Silencing by Antisense Genes

Antisense genes have been successfully used to alter plant process such as flower pigmentation, fruit ripening and photosynthesis and to determine the function of cryptic genes (for a detailed description of the applications of antisense genes in plants see GRIERSON et al. 1990 and VAN BLOKLAND et al. 1993). Besides improving fundamental knowledge of biological processes, the antisense approach has been applied in the field of biotechnology, for instance to increase the shelf-life of fruit (FRAY and GRIERSON 1993).

2.1.1 Modification of Flower Pigmentation in *Petunia hybrida*

The expression of several genes involved in flavonoid biosynthesis has been successfully modified by antisense genes. VAN DER KROL et al. (1988 b) were the first to show inhibition of gene expression by antisense constructs in transgenic plants. Chalcone synthase (CHS) is the key enzyme in flavonoid biosynthesis. The genes for *chs* were previously cloned by KOES et al. (1989). These authors showed that *chs* genes comprise a multigene family of which the members *chsA* and *chsJ* are transcribed. Petunia plants of the variety VR were transformed with the cDNA of *chsA*, fused in the antisense orientation to the CaMV-35S promoter. Thirteen out of 25 petunia VR transformations had an altered flower coloration ranging from purple to fully white. Some plants had white flowers with a coloured edge whereas others had flowers containing white sectors like the petunia variety Red Star (MOL et al. 1983). In unpigmented tissues, the *chs* mRNA levels were reduced by more than 90%, whereas transcript levels of others genes of the flavonoid pathway were unaffected (VAN DER KROL et al. 1988b). In the inbred purple line V26 only 10% of the transformants showed an altered phenotype. The phenotypes were, however, comparable to those obtained in the VR hybrid (VAN BLOKLAND 1994). To investigate whether other genes of the flavonoid pathway could be inactivated via a similar approach, the cDNA of dihydroflavonol 4-reductase (*dfr*) was fused in the antisense orientation to the 35S promoter and introduced into petunia V26 plants. The phenotypes obtained were comparable to those of antisense *chs* transformants. Six out of 18 plants showed an altered phenotype (VAN BLOKLAND 1994). As was the case with *chs*, the *dfr* transcripts were virtually absent in white corolla tissue. It is striking that using *chs* and *dfr* antisense genes a spectrum of pigmentation patterns was observed. Even the flowers from a single transgenic plant showed considerable variation in pigmentation. These results suggest that even though the antisense transcripts are produced, the effect of antisense RNA is not the same throughout the tissue. The pigmentation patterns observed in corollas of transformations containing the antisense *chs* genes change in response to the phytohormone gibberellin. Exogenous application of gibberellic acid (GA_3) or B9, an inhibitor of GA synthesis, increased and decreased the pigmented area, respectively. In addition, the size of the pigmented area is influenced by light during flower development. A high light intensity leads to an increase in the number of unpigmented cells, as is also the case in the natural Red Star variety (VAN DER KROL et al. 1990b). These findings indicate that antisense genes are effective in manipulating flower coloration but that it is influenced by the growth conditions.

2.1.2 Determining Gene Function

Differential screening or PCR-based strategies which are commonly used to isolate genes often result in a collection of cryptic genes. Expressing those genes in the sense or antisense orientation will provide information about their function provided that an altered plant phenotype is observed. By differential screening of flower-specific cDNA libraries and by expressing cDNA clones in the antisense orientation, one cDNA was shown to be involved in the 3-rhamnosylation of flavonoids (KROON et al. 1994; BRUGLIERA et al. 1994). Using the same approach, cDNAs isolated from a tomato ripening-specific library, pTOM13 and pTOM5 have been shown to encode ACC oxidase and the carotenoid-specific enzyme prophytoene synthase, respectively (HAMILTON et al 1990; BIRD et al. 1991).

Inhibition of the expression of the flavonoid gene *chs* in anthers of petunia leads to the production of white pollen. This flavonoid-free pollen is self-sterile and failed to produce pollen tubes in vitro. External application of flavonols could overcome this defect. Thus these antisense experiments have revealed a novel function for flavonoids (MO et al. 1992; VAN DER MEER et al. 1992; YLSTRA et al. 1992).

2.2 Antisense RNA May Block mRNA Production in More than One Way

In the presence of antisense genes complementary mRNA levels can be reduced. It is generally assumed that antisense RNA forms a dsRNA with its complementary mRNA leading to mRNA degradation, inhibition of RNA processing, transport or translation. Here we will discuss these possibilities for the action of the antisense flavonoid genes.

2.2.1 Interference of Antisense RNA with mRNA Processing and Transport

Some antisense RNAs have been shown to prevent processing of pre-mRNAs in vitro by masking sequences recognized by the splicing and the polyadenylation apparatus. We have indeed observed accumulation of unspliced *chs* RNAs in nuclear RNA preparations from flowers in which the *chs* gene was suppressed (VAN BLOKLAND 1994). However, transgenes and endogenous genes which do not contain introns can also be suppressed (ROTHSTEIN et al. 1987; DELAUNEY et al. 1988; CORNELISSEN and VANDEWIELE 1989; ROBERT et al. 1989; DAY et al. 1991; KROON et al. 1994). Even though it cannot be excluded that in some cases antisense RNAs inhibit splicing, it does not seem to be a major route of antisense RNA action. Accumulation of unspliced RNAs may therefore be a consequence of another event involving antisense RNA rather that the cause of gene suppression.

One may speculate that antisense RNAs disturb normal transport of complementary mRNAs out of the nucleus. Although it is conceivable that dsRNAs are not exported, as far as we know there are no experimental data to support this possibility.

2.2.2 Inhibition of Translation

Only two cases of translation inhibition of plant genes by antisense genes have been reported (CORNELISSEN and VANDEWIELE 1989; TEMPLE et al. 1993). In general only a fraction (10%–50%) of the transgenic plants expressing a particular antisense gene show suppression of the homologous gene. In extreme cases antisense genes may have no effect at all. For example, more than 60 petunia transformants have been generated expressing an antisense chalcone flavanone isomerase (*chi*) transgene. None of them show a shift in flower colour or a decrease in *chi* mRNA levels. Most transformants contain an excess antisense *chi* RNA. This clearly indicates that two complementary RNAs can coexist in the cytoplasm without obstructing the translation machinery.

2.2.3 dsRNA Formation and RNA Degradation

In many cases antisense genes prevent that accumulation of the target mRNA. Although it cannot be excluded that antisense RNA may interact with the target mRNA by blocking transcription, pre-mRNA processing or translation (see the previous sections), several studies in bacterial (CESARENI et al. 1991), fungal (NELLEN and LICHTENSTEIN 1993) and mammalian systems (KIM and WOLD 1985; BUNCH and GOLDSTEIN 1989; KIMELMAN and KIRSCHNER 1989; VOLK et al. 1989; KRYSTAL et al. 1990) indicate that the mRNA and complementary antisense RNA are either stably maintained in the cell or are simultaneously degraded. The simplest scenario for the latter phenomenon is that the sense and antisense RNAs form a dsRNA intermediate that is rapidly degraded by dsRNA-specific ribonucleases.

In a recent paper, NELLEN and LICHTENSTEIN (1993) discussed the issue of "antisense-itiveness" of mRNA. They reviewed the evidence for proteins that promote or inhibit RNA-RNA annealing and enzymes that specifically degrade dsRNAs. The authors stress that the in vitro conditions optimal for RNA-RNA annealing (50% formamide, 57°C, RNA probe in excess) are far removed from those in the cell. Therefore it seems likely that rapid in vivo dsRNA formation requires the help of a hybrid-promoting protein. A likely candidate is the RNP A1, which exhibits two functions in vitro: it binds aspecifically to single-stranded RNA (ssRNA) and it enhances the hybridization efficiency. Other proteins such as the unwindase/modificase (Bass and Weintraub 1988) or the DEAD box family of RNA helicases resolve dsRNA molecules. The first class or proteins seems to generate a paradox: resolving of RNA duplexes should lead to non-functional antisensing. However, unwinding is accompanied by inosination of RNAs, which would no longer encode functional proteins. The helicases on the other hand would just prevent functional "antisensing".

Attempts to identify duplexes between mRNAs and antisense RNAs in plants have been unsuccessful thus far (R. van Blokland, unpublished results and D. Grierson, personal communication). Indirect evidence for the involvement of antisense RNA in the degradation of its complementary mRNA in plants comes from studies where the antisense RNAs can be detected only in the absence of mRNA production. For example, in leaves of transgenic petunia plants in which the *chs* genes are expressed at low levels, antisense *chs* transcripts can be quite abundant (van der Krol et at. 1988b), whereas in flowers in which the *chs* gene is expressed, antisense *chs* transcripts are barely detectable. In transgenic petunia plants containing CaMV 35S promoter-driven antisense *chs* genes, antisense *chs* transcripts are present at relatively high levels, but they are barely detectable in flowers in which the expression of the resident *chs* gene is inhibited. A comparable antisense RNA expression pattern has been found for a tomato ripening-specific gene (Smith et al. 1990b).

Limited data are available on the involvement of dsRNA-specific nucleases in the turnover of dsRNAs. In prokaryotes, dsRNAses such as RNaseIII (Robertson et al. 1968; Robertson 1982) play an essential role in antisense RNA-regulated processes (Krinke and Wulff 1990). An RNAseIII-like activity has also been inferred in the degradation of the bFGF dsRNA in *Xenopus* oocytes (Kimelman and Kirschner 1989). Nellen has identified an activity in the slime mold *Dictyostelium* that specifically degrades the double-stranded portion from RNA hybrids (Nellen and Lichtenstein 1993). However, such an activity has not yet been demonstrated in plants.

2.3 Factors Involved in the Efficiency of Antisense Flavonoid Genes

2.3.1 The Amount of Antisense RNA

Based on kinetic considerations one would expect high concentrations of antisense RNA to increase the efficiency of suppression. In general this is indeed the case. In theory, high concentrations of antisense RNA may be achieved by using strong promoters, by insertion of multiple gene copies or by stabilization of the antisense RNA. Several studies have shown that antisense RNA is only effective if it is produced using strong promoters which are active in the same cell types as those of the resident genes (Robert et al. 1989; Cannon et al. 1990; van der Meer et al. 1992). A positive effect of multiple gene copies has been demonstrated by Smith et al. (1990a). Self-pollination of a tomato transformant containing a single antisense *pg* gene locus, yielded progeny containing either zero, one or two copies of the locus, which showed 100%, 20% and 1% PG activity, respectively. Similarly data have been reported for genes encoding ethylene-forming enzyme (EFE) in tomato (HAMILTON et al. 1990). We are not aware of any case in which the efficiency of an antisense RNA has been increased by incorporating RNA-stabilizing sequences (SULLIVAN and GREEN 1993).

There is not always a clear correlation between the steady-state level of antisense RNA and the inhibitory effects on mRNA accumulation (see for example VAN DER KROL et al. 1988b; CANNON et al. 1990; STOCKHAUS et al. 1990). To illustrate this further, the transcription rate of the antisense *chs* gene in petunia flowers was determined by nuclear run-on transcription analysis. In some transformants the antisense gene was highly transcribed in the flowers, yet did not lead to *chs* suppression, whereas in others the opposite was found (VAN BLOKLAND 1994). Even a promoterless *chs* gene induced phenotypic effects in 15% of the transformants VAN BLOKLAND et al. (1994). Nuclear run-on assays indicated that the *chs* transgene was not transcribed to a significant level in either direction, whereas the transcription rate of the resident *chs* genes was unaffected. These data show that a virtually silent transgene is able to suppress endogenous genes by a post-transcriptional process. Another puzzling observation is the local antisense effect giving rise to white sectors in flowers. Although the antisense genes were found to be transcribed throughout the corolla, the antisense RNAs appeared functional only in particular areas. The size of the pigmented sectors could be influenced by the exogenous application of hormones or by modifying light conditions, as has been described in a previous section. From these observations it is concluded that physiological conditions can influence the efficiency of antisense genes, and that the mere presence of antisense RNA does not guarantee suppression. The latter point is further illustrated by the inability of a *chi* antisense gene to suppress CHI activity in spite of the abundant production of antisense RNA in some of the transformants (see below).

2.3.2 Nucleotide Sequence Homology Between Resident and Antisense Genes

It would be convenient if a single antisense gene construct could be used to downregulate multiple copies of genes belonging to a multigene family or to suppress the homologous gene(s) in a variety of plant species. For example, *chs* genes in different lines of petunia comprise a multigene family of 8–12 members of which only two, *chsA* and *chsJ*, are transcribed in the flower (KOES et al. 1989). The nucleotide sequences of *chsA* and *J* differ by 15%, but an antisense *chsA* gene inhibits the expression of both *chs* genes. Moreover, the petunia *chsA* antisense gene also inhibits *chs* expression in the flowers of potato and tobacco (MOL et al. 1990; VAN DER KROL et al. 1988a). The similarity between these *chs* genes is not known, but might be lower than 80%. The overall sequence homology could be low, however, since relatively short patches of highly homologous sequences appear to be sufficient for a successful antisense effect (VISSER 1989; TEMPLE et al. 1993).

2.3.3 Antisense Gene Fragments

The results obtained with different antisense genes in various systems indicate that antisense RNAs may prevent mRNA accumulation following different pathways. Each transcript may have its own requirements for being attacked by antisense RNA. It is even conceivable that different parts of an antisense gene suppress gene expression with different efficiencies. VAN DER KROL et al. (1990c) identified several active antisense fragments complementary to different portions of the *chs* gene, the shortest being a 3'-fragment of 157 bp. As mentioned previously, the *chi* gene of petunia could not be inactivated by full-length antisense *chi* transgenes (MOL et al. 1990). Even antisense gene fragments complementary to the 3'-part (241 bp; 442 bp), the middle part 245 bp) and the 5'-part of the *chi* gene (284 bp) were completely inactive (DE LANGE 1994).

Other groups have identified active fragments complementary to different parts of the *nos* gene (ROTHSTEIN et al. 1987; SANDLER et al. 1988) and the β-glucuronidase (*uidA*) gene (CANNON et al. 1990; DE LANGE et al. 1993). However, the available data do not allow a rational design of efficient antisense gene fragments. "Antisensing" is still a matter of trial and error. Possible important parameters that need further study are the size, sequence, secondary structure and stability of antisense RNA fragments.

2.3.4 Is Genomic Position an Important Parameter?

Only 10%–50% of the transgenic petunia plants that carry a transcribed antisense flavonoid gene show suppression of the corresponding endogenous gene. As argued previously, this observation cannot be explained by transcription level differences alone. Thus, even though an excess of antisense RNA may be present in some cases, it does not necessarily reduce the level of complementary mRNA. If we assume that antisense RNAs work most efficiently in the nucleus, it is conceivable that the relative position of antisense genes and their targets in the interphase nucleus determines the interaction efficiency of the sense and antisense transcripts. RNAs synthesized in distant nuclear compartments may not sense each other and therefore fail to interact. We are not aware of any published data addressing this issue.

Introduction of genes into plant cells via *Agrobacterium tumefaciens* results in several independent transformants in which the transgenes reside at different chromosomal positions (HOOYKAAS 1989). Different antisense transgenes are expected to be similarly distributed throughout the plant genome. This may indicate that the failure to suppress the *chiA* gene is related to the genomic position of the resident *chiA* gene of petunia. To test this idea, we studied the interaction between sense and antisense *chi* transgenes. The petunia line W115 expresses the *chiA* gene in flower limbs but not in anthers because of a mutation in the promoter of the *chiA* gene. As a consequence the anthers accumulate yellow naringenin chalcone (VAN TUNEN et al. 1989). This line was transformed with a 35S promoter driven-sense *chiA* gene. One out of six transgenic plants had white anthers, which indicated the functional complementation of the *chi* mutant phenotype (VAN TUNEN et al. 1989). An antisense *chiA* gene fragment complementary to the 3'-end of the target gene was introduced into this transformant by a cross with a transformant carrying this 3'-end antisense fragment. Anthers of the F1 hybrid showed yellow pigmentation again, as in wild-type W115, indicating that the antisense and sense *chi* transgenes had interacted in some way. In corolla tissue, which expresses the *chiA* gene, the level of the RNA transcribed from the sense *chi* transgenes was drastically reduced, in contrast to that of the RNAs derived from the native *chiA* gene. By nuclear run-ons we could show that the transcription rates of both sense and antisense *chi* transgenes as compared to those in the parents were not reduced. This led us to conclude that the sense *chi* transgene was inactivated by an antisense *chi* transgene in a similar manner as described above for *chs* endogenous genes. In some of the other transformants uni- or bidirectional transcriptional silencing of the different *chi* transgenes was observed, which was associated with transgene methylation. However, in none of the transformants in which expression of the *chi* transgenes was affected, was expression of the endogenous *chiA* gene suppressed. We favour the idea that this differential sensitivity to suppression is related to the different genomic positions.

2.4 Natural Antisense RNAs in Plants

The occurrence of natural antisense RNAs has been described in bacteria, fungi and animals (for a review, see van Blokland et al. 1993). In only a few cases has natural antisense RNA been observed in plants. For example, complementary transcripts to the α-amylase mRNA are present in barley (Rogers 1988). The presence of antisense α-tubulin transcripts in *Zea mays* has recently been described by Dolfini et al. (1993). Schmitz and Theres (1992) have described the occurrence of overlapping transcripts derived from the *Bz2* locus of maize. This locus is involved in one of the later steps in anthocyanin biosynthesis. Since *Bz2* is a single-copy gene, the overlapping transcripts must be derived from the same locus. The biological function of all these natural antisense RNAs, if any, remains to be determined. Low amounts of antisense transcripts complementary to the *chs* genes have been detected in the petunia varieties Red Star and Velvet Picotee (van Blokland 1994). These varieties exhibit phenotypes similar to those produced by antisense *chs* transgenes and it is therefore tempting to speculate that the presence of these natural antisense transcripts is related to the formation the flower colour pattern. How these antisense transcripts are generated is unclear at present.

3 Gene Silencing by Sense Genes

Overexpression of genes is one of the strategies employed to overproduce specific proteins or to alter fluxes in metabolic pathways. To achieve this, cloned genes transcribed by strong constitutive or tissue-specific promoters are transferred into plants and suitable transformants are selected. However, in contrast to expectations, gene suppression is found in a considerable proportion of the plants. In these cases, either the resident gene, the introduced transgenes or both have been found to be inactivated. This phenomenon is termed co-suppression or sense suppression (JORGENSEN 1992).

3.1 Examples of Gene Silencing by Sense Genes

Several cases of gene silencing in plants by sense genes have been documented whereby distinct types of gene suppression could be distinguished: transgene inactivation and methylation in multicopy transformants, transgene silencing in sequentially transformed plants and post-transcriptional gene suppression (for reviews, see JORGENSEN 1990, 1992; GRIERSON et al. 1991; KOOTER and MOL 1993; MATZKE and MATZKE 1993; MEYER et al. 1993; VAN BLOKLAND et al. 1993; MOL et al. 1994). It is not clear to what extent the results obtained with the different systems can be compared because the mechanisms could be different. However, a key factor in all these cases seems to be the sequence homology between the suppressed and the suppressing genes, as if some kind of homology-sensing machinery is operating.

3.1.1 Transgene Inactivation and Methylation in Multicopy Transformants

Transcriptional inhibition of transgenes is very often correlated with increased methylation (LINN et al. 1990; ASSAAD et al. 1993; HOBBS et al. 1993; MATZKE et al. 1993). Mutual suppression of transgenes sometimes occurs in plants that carry multiple copies of the transgenes. This phenomenon has been observed with dihydroflavonol 4-reductase (*dfr*) transgenes in petunia (MEYER et al. 1987; LINN et al. 1990), *uidA* genes in tobacco (HOBBS et al. 1993) and hygromycin (*hpt*) and neomycin (*npt*) genes in *Arabidopsis* (MITTELSTEN SCHEID et al. 1991; ASSAAD et al. 1993). Some of these cases are described elsewhere in this volume.

3.1.2 Post-transcriptional Suppression of Chalcone Synthase Genes

Not all gene silencing phenomena caused by transgenes can be explained by methylation of the interacting genes. Dosage-dependent suppression of a β-1,3-glucanase transgene (DE CARVALHO et al. 1992), as well as suppression of phenylalanine ammonia-lyase (ELKIND et al. 1990; BATE et al. 1992) and CHS (MOL et al. 1991; VAN BLOKLAND 1994; VAN BLOKLAND et al. 1994) by homologous transgenes are clear examples of post-transcriptional processes.

Mutual silencing of transgenes and endogenous genes was first reported by NAPOLI et al. (1990) and VAN DER KROL et al. (1990a). Transfer of genes from the flower pigmentation pathway encoding the enzymes CHS and dihydroflavonol 4-reductase back into petunia resulted in transformants with reduced flower pigmentation. Some plants produced completely white flowers, whereas others showed pigmentation patterns. These altered phenotypes were observed in

20%–50% of the transformants. In the white sectors of the corolla of *chs* sense transformants both the endogenous and the transgene *chs* RNAs were strongly reduced, but analysis of nascent *chs* RNAs revealed that the flavonoid genes were transcribed at normal rates and *chs* RNAs correctly processed and polyadenylated (VAN BLOKLAND 1994). Interestingly, a slight accumulation of unspliced *chs* precursor RNAs was found. However, spliced transcripts were also detectable (although at lower levels than in wild-type tissue), indicating that splicing is not completely blocked. Moreover, intronless transgenes are suppressed simultaneously with native intron-containing genes. We therefore think that the seemingly reduced splicing is a consequence of another process which is affected by the presumed interaction, rather than itself being the cause of *chs* suppression.

Sense suppression is not caused by transcriptional silencing but rather by preventing accumulation of mature mRNAs. The strength of the promoter used to drive transcription of the transgenes seems not to be of critical importance. Both the CaMV-35S and the endogenous *chs* promoter have proved effective (VAN DER KROL et al. 1990a); even promoterless *chs* genes gave phenotypic effects in 15% of transgenic petunia plants (VAN BLOKLAND 1994; VAN BLOKLAND et al. 1994). Pigmentation patterns are influenced by light and hormones in a similar manner as is observed with antisense transformants (VAN DER KROL et al. 1990b; M. Stam, unpublished results). There is no positive correlation between a particular rate of transgene transcription and *chs* suppression. In the case of local *chs* suppression the transcription rates of the *chs* genes are identical in pigmented and unpigmented sectors. Taken together, the data obtained so far in this system indicate that the action of extra sense genes resembles that of antisense RNA producing genes. Silencing of homologous plant genes has been observed with chimeric (*uidA-chsA*; *uidA-dfr*) and truncated *chs* transgenes (VAN BLOKLAND 1994) as well as with truncated polygalacturonidase (*pg*) genes (SMITH et al. 1990b). As in these cases no functional CHS or PG enzyme is made, mechanisms involving feedback inhibition by elevated levels of gene products or metabolites are unlikely.

3.2 Gene Suppression Without Transgenes

A number of cases where gene suppression occurs in the absence of transgenes have been documented. For example, paramutation is an epigenetic phenomenon in which a directed alteration of one allele is caused by the presence of another allele of the gene. PATTERSON et al. (1993) have studied paramutation of alleles of the *b* gene involved in anthocyanin pigmentation. Plants homozygous for the strong *B-I* allele produce high levels of anthocyanins. *B′* is a weak spontaneous mutant allele of *B-I*; plants homozygous for *B′* produce little anthocyanin. *B′/B-I* heterozygotes, which are expected to produce an intermediate amount of anthocyanin, behave as *B′/B′* homozygotes. Apparently, *B′* is dominant over *B-I*. The *B-I* allele is never transmitted from a *B′/B-I* heterozygote; it is changed to *B′*. The change from *B-I* to *B′* is accompanied by a drastic reduction in transcription. However, Patterson et al. found no evidence for rearrangements or changes in sequence or in methylation status. They suggested that *b* paramutation involves a physical interaction between alleles that promotes a heritable alteration in chromatin structure (see also JORGENSEN 1992).

Several semidominant alleles have been described for the *nivea* locus of *Antirrhinum majus* (encoding CHS) which resemble introduced sense genes in their ability to suppress the activity of the wild-type *chs* allele *in trans* (BOLLMANN et al. 1991). These *chs* alleles were derived from the unstable *niv*-93 allele, which has the Tam3 transposon inserted into the promoter (COEN and CARPENTER 1986). Structural analysis of three semidominant *nivea* alleles showed the presence of inversions and multiple copies of *niv* gene sequences (BOLLMANN et al. 1991). Although the cause of the semidominance of these *nivea* alleles is not known, the duplication/rearrangement process may have features in common with the T-DNA integration event. Other examples of natural suppression of *chs* genes have been found in the petunia varieties Red Star and Velvet Picotee. Flowers of the first variety have alternating red and white sectors whereas Velvet Picotee has a white ring at the outer edge of the corolla and a coloured tube. The *chs* mRNA level in white sectors is severely reduced as compared to the red sectors (MOL et al. 1983; VAN BLOKLAND 1994). However, in the white sectors, *chs* is transcribed at the same rate as in the pigmented parts. This is in agreement with the observation that the *chs* promoter directs expression of the *cat* reporter gene equally well in coloured and unpigmented sectors of the flowers of transgenic Red Star (VAN DER MEER 1991). These data indicate that in Red Star and Velvet Picotee the expression of *chs* is also inhibited by a post-transcriptional mechanism. In both varieties (but not in wild-type petunia) a low level of antisense *chs* gene transcription was detected (VAN BLOKLAND 1994). It is tempting to speculate that one member of the *chs* gene family is rearranged analogous to the semidominant *nivea* locus and mimics an antisense *chs* transgene. It remains unclear, however, how these rearrangements generate such regular pigmentation patterns in flowers. Since both the abaxial and adaxial side of the corolla are affected in a similar manner, the pigmentation pattern might be established early during flower development.

4 Are There Different Mechanisms for Antisense and Sense Gene-Mediated Silencing?

A number of different cases have been described in which homologous antisense and sense genes suppress gene expression in plants. The mechanism(s) by which antisense genes work in plants is still unclear. In those cases where a strong correlation is observed between gene dosage and gene suppression (HAMILTON et al. 1990; SMITH et al. 1990a), the mechanism may involve the formation of dsRNA. However, attempts to identify duplexes between antisense RNA and mRNA in plants have been unsuccessful so far. In cases where small amounts of antisense RNA are effective or where local effects are observed (CANNON et al. 1990; STOCKHAUS et al. 1990; VAN BLOKLAND 1994), additional factors must come into play. In other cases where sense transgenes interact with antisense transgenes, transcriptional silencing can occur independent of the production of antisense RNA (YIBRAH et al. 1993; DE LANGE 1994).

Several mechanisms have been proposed in the literature to explain sense gene-induced suppression, some of which are based on similar phenomena in other organisms (JORGENSEN 1990, 1992). They include:

1. inactivation of genes by methylation induced by the presence of homologous sequences in trans (LINN et al. 1990; HOBBS et al. 1993; MATZKE et al. 1993) or in cis by repeated sequences (ASSAAD et al. 1993). The latter resembles the mechanism of repeat-induced point mutation and methylation induced pre-meiotically in filamentous fungi (CAMBARERI et al. 1989; SELKER 1990, 1991; RHOUNIM et al. 1992).
2. The production of "unintended" antisense RNA (GRIERSON et al. 1991).
3. Allelic interactions proposed to occur in the case of paramutation in plants (BRINK 1973; PATTERSON et al. 1993). Such allelic interactions can be envisaged by pairing of homologous loci or chromosomes. In the case of transvection in Drosophila the allelic interaction (TARTOF and HENIKOFF 1991) is mediated by DNA-binding proteins that interact with promoter sequences (CHEN et al. 1992) and results in a transcriptional activation. In the other cases the mechanism is unknown. It is conceivable that sequence homology between chromosomal loci, whether allelic or ectopic, is sufficient for pairing although such a suggestion is speculative, pairing could be further enhanced by transcripts, especially if one of the loci produces antisense RNA.

4.1 Could There Be a Role for Antisense RNA in Plant Gene Suppression by Sense Genes?

Since sense suppression has so many features in common with antisense suppression, one may speculate that the mechanism of action of antisense and sense genes share common steps.

It has been suggested that sense suppression is caused by the production of "unintended" antisense RNA molecules. The ways in which antisense RNA could be transcribed from an introduced sense transgene are discussed by GRIERSON et al. (1991). This idea is supported by the observation that in petunia plants transformed with a sense *chs* gene, the production of very small amounts of nascent antisense *chs* transcripts could be detected (VAN BLOKLAND 1994). The production of nascent antisense *chs* transcripts was also observed in transformants in which flower coloration was normal or in sectors. In untransformed controls antisense *chs* transcripts levels were below our detection limits. Even in plants harbouring promoterless *chs* cDNAs, minute amounts of antisense *chs* transcripts were produced, both in white and in pigmented parts of the corolla.

The *chs* mutants Red Star and Velvet Picotee also produce small amounts of antisense *chs* transcripts (VAN BLOKLAND et al. 1994).

An attractive feature of the antisense RNA model is that it explains the remarkable similarity in the effects of sense and antisense *chs* genes. However, since antisense RNA production is at a much lower level than sense and endogenous RNA production in the sense transformants, data are hard to accommodate within a model in which a sense RNA interacts directly with an antisense RNA since this would require stoichiometric amounts of antisense RNA. If we assume that a dsRNA is an intermediate in the degradation process, it is difficult to understand how an excess of mRNA can be degraded, unless one assumes that antisense RNA can work catalytically or that antisense RNAs can hybridize simultaneously to multiple nascent mRNA molecules by virtue of the proximity of the genetic loci which produce these RNAs.

4.2 Allelic and Ectopic Interactions

In those cases where antisense RNA cannot easily account for gene silencing, e.g. the local effects of sense and antisense *chs* genes, alternative explanations should be provided. Titration of transcription factors is a possibility; however, this would only account for those cases where transcriptional silencing occurs and where a correlation between the extent of gene silencing and the number of expressed gene copies is evident. Another possibility is specific feedback regulation by elevated levels of gene products or metabolites. This type of explanation can only account for the reported silencing of β-1,3-glucanase and *pal* genes and not for silencing of *chs* and *pg* expression by truncated genes or of that by foreign genes. A recently advanced alternative to explain sense gene-induced silencing is ectopic pairing (JORGENSEN 1990, 1992). In this model a direct interaction between homologous DNA sequences may lead to exchange of (unspecified) chromation components. In this way two distinct allelic or ectopic gene pairs may be "harmonized", the result of which may be silencing. The silencing principle is unknown so far, but heterochromatization may be one of the possibilities (ASSAAD et al. 1993).

In a normal situation unperturbed allelic gene pairs or gene family members are not silenced. Thus, sequence homology per se is not sufficient to achieve gene silencing. Evidence is accumulating that the structure of (one of) the homologous loci, whether ectopic or allelic, may determine whether they interact in a transcriptional/post-transcriptional mode or not at all. For example, HOBBS et al. (1993) found that transcriptional silencing of *uidA* transgenes was correlated with the presence of T-DNA inverted repeats. ASSAAD et al. (1993) identified expressed single-copy and silenced multicopy inserts at the same locus. Phenotypic effects of *chs* transgenes in petunia are correlated with the presence of inverted repeats of T-DNA (VAN BLOKLAND 1994). Because of the local effects in some plants carrying *chs* transgenes, the structure of integrated T-DNAs cannot be the only factor involved.

5 Conclusions and Future Prospects

The use of sense and antisense genes has provided a wealth of information on the regulation of primary and secondary metabolism in plants and facilitated the identification of cryptic genes. The availability of cell-type and developmental stage-specific promoters allows the spatial and temporal suppression of gene expression. Transformation of plants with a specific construct yields a collection of transformants with various expression levels and phenotypes. Such a collection of plants cannot be obtained easily by other approaches such as radiation, T-DNA and transposon mutagenesis or homologous recombination. Therefore, modulation of gene expression can be best achieved by sense and antisense genes.

A number of problems remain to be solved, however. For instance, the genetic stability of transgenes cannot be controlled over many generations. This problem has to be solved if crop production on a commercial basis is aimed at. The use of scaffold attachment sites may improve the genetic stability of the transgenes by eliminating effects of flanking sequences (STIEF et al. 1989; PHI VAN et al. 1990; BREYNE et al. 1992; SCHÖFFL et al. 1993). Another potential problem is that one cannot predict the efficiency of sense and antisense genes on the basis of their size and sequence. If we understand the mechanism of sense and antisense suppression this issue may be resolved; however, the situation we face today is still rather complicated and confusing since some combinations of genes interact at the transcriptional level whereas others do so post-transcriptionally. The same combinations of sense and antisense genes can even display different modes of gene silencing (YIBRAH et al. 1993; this chapter). In the cases studied, gene methylation is clearly a function of transgene dosage and only affects transgenes (LINN et al. 1990; HOBBS et al. 1993). We are not aware of any case whereby a transgene causes methylation of a resident plant gene. The post-transcriptional mode of action of some antisense and sense genes is an even greater mystery. The action, for example, of antisense and sense *chs* genes cannot readily be explained by a simple model involving duplex RNA formation. Theoretically, low production of antisense RNA may be sufficient to obtain post-transcriptional silencing of the homologous resident copy, e.g. by facilitating pairing of ectopic gene copies. It is not known, however, whether antisense RNA can accomplish this. For other genes such as *efe* and *pg* in tomato, for which suppression depends on gene dosage, silencing could operate through duplex RNA formation. Whereas the production of "unintended" antisense RNA may provide a plausible explanation for the observed effects of sense *chs* transgenes, it cannot readily explain the suppression of β-1,3-glucanase and *pal* genes. Suppression of β-1,3-glucanase and *pal* might have been caused by a biochemical switch which is activated by increased levels of enzymes or metabolites. Gene silencing by transgenes seems to depend on the nucleotide sequence homology of the loci involved. It is conceivable that chromatin structure and the relative location of genes in the nucleus play a key role in this process. The use of in situ hybridization techniques and the analysis of chromatin structure may provide further insights into such parameters.

Whether co-suppression occurs in the nucleus or in the cytoplasm is unknown. Support for a nuclear process could come from experiments with sense and antisense intron sequences. To our knowledge such experiments have not been reported in the literature. Post-transcriptional suppression of genes may occur outside the nucleus (LINDBO et al. 1993). These authors have shown that accumulation of RNA of the tobacco etch virus (TEV) in infected tobacco plants can be inhibited by introduction of a TEV gene encoding a truncated, non-functional coat protein. The transgene is transcribed normally but steady-state transcript levels are severely reduced. Whether this phenomenon is related to the other cases of gene silencing remains to be determined.

In conclusion, we believe that further analysis of nuclear architecture, chromatin structure and determinants of RNA stability will aid our efforts to understand the different mechanisms of gene silencing in plants.

References

Assaad FF, Tucker KL, Signer ER (1993) Epigenetic repeat-induced gene silencing (RIGS) in Arabidopsis. Plant Mol Biol 22: 1067–1085

Bass BL, Weintraub H (1987) A developmentally regulated activity that unwinds RNA duplexes. Cell 48: 607–613

Bass BL, Weintraub H (1988) An unwinding activity that covalently modifies its double-stranded RNA substrates. Cell 55:1089–1098

Bate NJ, Pallas JA, Elkind Y, Lamb CJ (1992) Sense-suppression of endogenous phenylalanine ammonia-lyase (PAL) expression in tobacco containing a bean PAL transgene. Abstract 9, the 10th John Innes symposium, the chromosome, Norwich, Sept 1992

Bird CR, Ray JA, Fletcher JD, Boniwell JM, Bird AS, Teulieres C, Blain I, Bramley PM, Schuch W (1991) Using antisense RNA to study gene function—inhibition of carotenoid biosynthesis in transgenic tomatoes. Biotechnology 9:635–639

Bollmann J, Carpenter R, Coen ES (1991) Allelic interactions at the nivea locus of Antirrhinum. Plant Cell 3: 1327–1336

Breyne P, Vanmontagu M, Depicker A, Gheysen G (1992) Characterization of a plant scaffold attachment region in a DNA fragment that normalizes transgene expression in tobacco. Plant Cell 4: 463–471

Brink RA (1973) Paramutation. Annu Rev Genet 7: 129–152

Brugliera F, Holton TA, Stevenson TW, Farcy E, Lu C-Y, Cornish EC (1994) Isolation and characterization of a cDNA clone corresponding to the Rt locus of Petunia hybrida. Plant J 5: 81–92

Bunch TA, Goldstein LSB (1989) The conditional inhibition of gene expression in cultured Drosophila cells by antisense RNA. Nucleic Acids Res 17: 9761–9781

Cambareri EB, Jensen BC, Schabtach E, Selker EU (1989) Repeat induced G-C to A-T mutations in Neurospora. Science 244: 1571–1575

Cannon M, Platz J, O'Leary M, Sookdeo C, Cannon F (1990) Organ-specific modulation of gene expression in transgenic plants using antisense RNA. Plant Mol Biol 15: 39–47

Cesareni G, Helmercitterich M, Castagnoli L (1991) Control of ColE1 plasmid replication by antisense RNA. Trends Genet 7: 230–235

Chen T, Bunting M, Karim FD, Thummel CS (1992) Isolation and characterization of 5 Drosophila genes that encode an ets-related DNA binding domain. Dev Biol 151: 176–191

Coen ES, Carpenter R (1986) Transposable elements in Antirrhinum majus: generators of genetic diversity. Trends Genet 20: 292–296

Cornelissen M, Vandewiele M (1989) Both RNA level and translation efficiency are reduced by antisense RNA in transgenic tobacco. Nucleic Acids Res 17: 833–843

Day AG, Bejarano ER, Buck KW, Burrell M, Lichtenstein CP (1991) Expression of an antisense viral gene in transgenic tobacco confers resistance to the DNA virus tomato golden mosaic virus. Proc Natl Acad Sci USA 88: 6721–6725

De Carvalho F, Gheysen G, Kushnir S, Van Montagu M, Inzé D, Castresana C (1992) Suppression of beta-1,3-glucanase transgene expression in homozygous plants. EMBO J 11 :2595–2602

De Lange P (1994) Inhibition of gene expression by antisense genes in Petunia hybrida. Thesis, Vrije Universiteit, Amsterdam, The Netherlands

De Lange P, De Boer G-J, Mol JNM, Kooter JM (1993) conditional inhibition of beta-glu-curonidase expression by antisense gene fragments in petunia protoplasts. Plant Mol Biol 23: 45–55

Delauney AJ, Tabaeizadeh Z, Verma DPS (1988) A stable bifunctional antisense transcript inhibiting gene expression in transgenic plants. Proc Natl Acad Sci USA 85: 4300–4304

Dolfini S, Consonni G, Mereghetti M, Tonelli C (1993) Antiparallel expression of the sense and antisense transcripts of maize alpha-tubulin genes. Mol Gen Genet 241: 161–169

Elkind Y, Edwards R, Mavandad M, Hedrick SA, Ribak O, Dixon RA, Lamb CJ (1990) Abnormal plant development and down-regulation of phenylpropanoid biosynthesis in transgenic tobacco containing a heterologous phenylalanine ammonia-lyase gene. Proc Natl Acad Sci USA 87: 9057–9061

Fray R, Grierson D (1993) Molecular genetics of tomato fruit ripening. Trends Genet 9: 438–443

Grierson D, Smith CJS, Watson CF, Morris PC, Turner AJ, Schuch W, Bird CR, Ray J, Hamilton A (1990) Controlling gene expression in transgenic tomatoes. Horticult Biotechnol 11: 229–236

Grierson D, Fray RG, Hamilton AJ, Smith CJS, Watson CF (1991) Does co-suppression of sense genes in transgenic plants involve antisense RNA? Trends Biotech 9: 122–123

Hamilton AJ, Lycett GW, Grierson D (1990) Antisense gene that inhibits synthesis of the hormone ethylene in transgenic plants. Nature 346: 284–287

Hobbs SLA, Warkentin TD, Delong CMO (1993) Transgene copy number can be positively or negatively associated with transgene expression. Plant Mol Biol 21: 17–26

Hooykaas P (1989) Transformation of plant cells via Agrobacterium. Plant Mol Biol 13: 327–336

Jorgensen R (1990) Altered gene expression in plants due to trans interactions between homologous genes. Trends Biotech 8: 340–344

Jorgensen R (1992) Silencing of plant genes by homologous transgenes. Agbiotech News Info 4: 265–273

Kim SK, Wold BJ (1985) Stable reduction of thymidine kinase activity in cells expressing high levels of anti-sense RNA. Cell 42: 129–138

Kimelman D, Kirschner MW (1989) An antisense mRNA directs the covalent modification of the transcript encoding fibroblast growth factor in Xenopus oocytes. Cell 59: 687–696

Koes RE, Spelt CE, Van den Elzen PJM, Mol JNM (1989) Cloning and molecular characterization of the chalcone synthase multigene family of Petunia hybrida. Gene 81: 245–257

Kooter JM, Mol JNM (1993) Trans-inactivation of gene expression in plants. Curr Opin Biotechnol 4: 166–171

Krinke L, Wulff DL (1990) RNase III-dependent hydrolysis of lambda-cII-O gene messenger RNA mediated by lambda OOP antisense RNA. Genes Dev 4: 2223–2233

Kroon J, Souer E, de Graaf A, Xue Y, Mol J, Koes R (1994) Cloning and structural analysis of the anthocyanin locus Rt of Petunia hybrida: characterization of insertion sequences in two mutant alleles. Plant J 5: 69–80

Krystal GW, Armstrong BC, Battey JF (1990) N-myc mRNA forms an RNA-RNA duplex with endogenous antisense transcripts. Mol Cell Biol 10: 4180–4191

Lindbo JA, Silva-Rosales L, Proebsting WM, Dougherty WG (1993) Induction of a highly specific antiviral state in transgenic plants: implications for regulation of gene expression and virus resistance. Plant Cell 5: 1749–1759

Linn F, Heidmann I, Saedler H, Meyer P (1990) Epigenetic changes in the expression of the maize A1 gene in Petunia hybrida: role of numbers of integrated gene copies and state of methylation. Mol Gen Genet 222: 329–336

Matzke M, Matzke AJM (1993) Genomic imprinting in plants–parental effects and trans-inactivation phenomena. Annu Rev Plant Physiol 44: 53–76

Matzke MA, Neuhuber F, Matzke AJM (1993) A variety of epistatic interactions can occur between partially homologous transgene loci brought together by sexual crossing. Mol Gen Genet 236: 379–386

Melton DA (1985) Injected anti-sense RNAs specifically block messenger RNA translation in vivo. Proc Natl Acad Sci USA 82: 144–148

Meyer P, Heidmann I, Forkmann G, Saedler H (1987) A new Petunia flower colour generated by transformation of a mutant with a maize gene. Nature 330: 677–678

Meyer P, Heidmann I, Niedenhof I (1993) Differences in DNA-methylation are associated with a paramutation phenomenon in transgenic petunia. Plant J 4: 89–100

Mittelsten Scheid OM, Paszkowski J, Potrykus I (1991) Reversible inactivation of a transgene in Arabidopsis thaliana. Mol Gen Genet 228; 104–112

Mo Y, Nagel C, Taylor LP (1992) Biochemical complementation of chalcone synthase mutants defines a role for flavonols in functional pollen. Proc Natl Acad Sci USA 89: 7213–7217

Mol J, Van Blokland R, Kooter J (1991) More about co-suppression. Trends Biotech 9: 182–183

Mol JNM, Schram AW, de Vlaming P, Gerats AGM, Kreuzaler F, Hahlbrock K, Reif HJ, Veltkamp E (1983) Regulation of flavonoid gene expression in Petunia hybrida: description and partial characterization of a conditional mutant in chalcone synthase gene expression. Mol Gen Genet 192: 424–429

Mol JNM, Van der Krol AR, Van Tunen AJ, Van Blokland R, De Lange P, Stuitje AR (1990) Regulation of plant gene expression by antisense RNA. FEBS Lett 268: 427–430

Mol JNM, Van Blockland R, de Lange P, Stam M, Kooter JM (1994) Post-transcriptional inhibition of gene expression: sense and antisense genes. In: Paszkowski J (ed) Homologous recombination in plants, Kluwer Academic, Dordrecht pp 309–334

Napoli C, Lemieux C, Jorgensen R (1990) Introduction of a chimeric chalcone synthase gene into petunia results in reversible co-suppression of homologous genes in trans. Plant Cell 2: 279–289

Nellen W, Lichtenstein C (1993) What makes an mRNA anti-sense-itive? Trends Biochem Sci 18: 419–423

Patterson GI, Thorpe CJ, Chandler VL (1993) Paramutation, an allelic interaction, is associated with a stable and heritable reduction of transcription of the maize-B regulatory gene. Genetics 135: 881–894

Phi-Van C, von Kries JP, Ostertag W, Strätling WH (1990) The chicken lysozyme 5' matrix attachment region increases transcription from a heterologous promoter in heterologous cells and dampens position effects on the expression of transfected genes, Mol Cell Biol 10: 2302–2307

Rhounim L, Rossignol J-L, Faugeron G (1992) Epimutation of repeated genes in Ascobolus immersus. EMBO J 11: 4451–4457

Robert LS, Donaldson PA, Ladaique C, Altosaar I, Arnison PG, Fabijanski SF (1989) Antisense RNA inhibition of beta-glucuronidase gene expression in transgenic tobacco plants. Plant Mol Biol 13: 399–409

Robertson HD (1982) Escherichia coli ribonuclease III cleavage sites. Cell 30: 669–672

Robertson HD, Webster RE, Zinder ND (1968) Purification and properties of ribonuclease III from Escherichia coli. J Biol Chem 243: 82–91

Rogers J (1988) RNA complementary to α-amylase mRNA in barley. Plant Mol Biol 11: 125–138

Rothstein PA, DiMaio J, Strand M, Rice D (1987) Stable and heritable inhibition of the expression of nopaline synthase in tobacco expressing anti-sense RNA. Proc Natl Acad Sci USA 84: 8439–8443

Sandler SJ, Stayton M, Townsend JA, Ralston ML, Bedbrook JR, Dunsmuir P (1988) Inhibition of gene expression in transformed plants by antisense RNA. Plant Mol Biol 11: 301–310

Schmitz G, Theres K (1992) Structural and functional analysis of the Bz2 locus in maize: characterization of overlapping transcripts. Mol Gen Genet 233: 269–277

Schöffl F, Schroder G, Kliem M, Rieping M (1993) An SAR-sequence containing 395 bp-DNA fragment mediates enhanced, gene-dosage-correlated expression of a chimaeric heat shock gene in transgenic tobacco plants. Transgenic Res 2: 93–100

Selker EU (1990) Premeiotic instability of repeated sequences in Neurospora crassa. Annu Rev Genet 24: 579–613

Selker EU (1991) Repeat-induced point mutation and DNA methylation. In: Bennett JW, Lasure LL (ed) More gene manipulations in fungi. Academic Press, San Diego, pp 258–265

Smith C, Watson CF, Morris PC, Bird CR, Seymour GB, Gray JE, Arnold C, Tucker GA, Schuch W, Harding S, Grierson D (1990a) Inheritance and effect on ripening of antisense polygalacturonase genes in transgenic tomatoes. Plant Mol Biol 14: 369–379

Smith CJS, Watson CF, Bird CR, Ray J, Schuch W, Grierson D (1990b) Expression of a truncated tomato polygalacturonase gene inhibits expression of the endogenous gene in plants. Mol Gen Genet 224: 477–481

Stief A, Winter DM, Strätling WH, Sippel AE (1989) A nuclear DNA attachment element mediates elevated and position-independent gene activity. Nature 341: 343–345

Stockhaus J, Hofer M, Renger G, Westhoff P, Wydrzynski T, Willmitzer L (1990) Anti-sense RNA efficiently inhibits formation of the 10 Kd polypeptide of photosystem-II in transgenic potato plants–analysis of the role of the 10 Kd protein. EMBO J 9: 3013–3021

Sullivan ML, Green PJ (1993) Post-transcriptional regulation of nuclear-encoded genes in higher plants: the roles of mRNA stability and translation. Plant Mol Biol 23: 1091–1104

Tartof KD, Henikoff S (1991) Trans-sensing effects from drosophila to humans. Cell 65: 201–203

Temple SJ, Knight TJ, Unkefer PJ, Senguptagopalan C (1993) Modulation of glutamine synthetase gene expression in tobacco by the introduction of an alfalfa glutamine synthetase gene in sense and antisense orientation—molecular and biochemical analysis. Mol Gen Genet 236: 315–325

Van Blokland R (1994) Trans-inactivation of flower pigmentation genes in Petunia hybrida. Thesis, Vrije Universiteit, Amsterdam, The Netherlands

van Blokland R, van der Geest N, Mol JNM, Kooter JM (1994) Transgene-mediated suppression of chalcone synthase expression in petunia hybrida results from an increase in RNA turnover. Plant J 6 (in press)

Van Blokland R, De Lange P, Mol JNM, Kooter JM (1993) Modulation of gene expression in plants by antisense genes. In: Lebleu B (ed) Antisense research and applications. CRC Press, Boca Raton, pp 125–148

Van der Krol A, Mol JNM, Stuitje AR (1988a) Modulation of eukaryotic gene expression by complementary RNA or DNA sequences. Biotechniques 6: 958–976

Van der Krol AR, Lenting PE, Veenstra J, Van der Meer IM, Koes RE, Gerats AGM, Mol JNM, Stuitje AR (1988b) An anti-sense chalcone synthase gene in transgenic plants inhibits flower pigmentation. Nature 333: 866–869

Van der Krol AR, Mur LA, Beld M, Mol JNM, Stuitje AR (1990a) Flavonoid genes in petunia: addition of a limited number of gene copies may lead to a suppression of gene expression. Plant Cell 2: 291–299

Van der Krol AR, Mur LA, de Lange P, Gerats AGM, Mol JNM, Stuitje AR (1990b) Antisense Chalcone synthase genes in Petunia: visualization of variable transgene expression. Mol Gen Genet 220: 204–212

Van der Krol AR, Mur LA, de Lange P, Mol JNM, Stuitje AR (1990c) Inhibition of flower pigmentation by antisense genes: promoter and minimal sequence requirements for the antisense effect. Plant Mol Biol 14: 457–466

Van der Meer IM (1991) Regulation of flavonoid gene expression in Petunia hybrida: cis-acting elements and trans-acting factors. Thesis, Vrije Universiteit Amsterdam

Van der Meer IM, Stam M, van Tunen AJ, Mol JNM, Stuitje AR (1992) Antisense inhibition of flavonoid biosynthesis in Petunia anthers results in male sterility. Plant Cell 4: 253–262

Van Tunen AJ, Hartman SA, Mur LA, Mol JNM (1989) Regulation of chalcone flavanone isomerase (CHI) gene expression in Petunia hybrida: the use of alternative promoters in corolla, anthers and pollen. Plant Mol Biol 12: 539–551

Visser RGF (1989) Manipulation of starch composition of Solanum tuberosum L. using Agrobacterium rhizogenes mediated transformation. Thesis, University of Groningen, Groningen, The Netherlands

Volk R, Köster M, Pötig A, Hartmann L, Knöchel W (1989) An antisense transcript from the Xenopus laevis bFGF gene coding for an evolutionarily conserved 24 kd protein. EMBO J 8: 2983–2988

Yibrah HS, Clapham DH, Vonarnold S (1993) Antisense RNA inhibition of uidA gene expression in transgenic plants–evidence for interaction between 1st and 2nd transformation events. Hereditas 118: 273–280

Ylstra B, Turaev A, Moreno BRM, Stöger E, Van Tunen AJ, Vicente O, Mol JNM, Heberle-Bors E (1992) Flavonols stimulate development, germination, and tube growth of tobacco pollen. Plant Physiol 100: 902–907

Sense and Antisense Inactivation of Fruit Ripening Genes in Tomato

A.J. Hamilton, R.G. Fray, and D. Grierson

1 Introduction

The tomato (*Lycopersicon esculentum*), and in particular the ripening fruit, has been the subject of numerous experiments involving antisense or sense transgene-mediated suppression of endogenous gene expression in plants, and commercial products developed with these techniques are likely to be on sale in 1995. The efficiency, reliability and technical simplicity of these genetic approaches belies our very limited understanding of how they work. This review discusses how experiments to modify tomato gene expression have provided a wealth of new physiological and biochemical information and the contribution they have made to our understanding of the nature of antisense and sense gene-mediated suppression.

BBSRC Research Group in Plant Gene Regulation, University of Nottingham, Sutton Bonington Campus, Loughborough, Leics LE 12 5RD, UK

2 Cell Wall Metabolism During Ripening

Plant cell walls are complex structures which give strength to plant organs, protect the enclosed cells from a variety of environmental insults and give shape to the cells, tissues and ultimately the plant as a whole. Therefore, precise control of cell wall structure, composition and biochemistry is of fundamental importance to the plant during growth and development. This is partly accomplished by the coordinated action of a number of hydrolytic enzymes which cleave the various polymers that constitute the cell wall (for review see McNeil et al. 1984 and Varner and Lin 1989). Cell wall changes which occur during fruit ripening have become an important model system for the study of the induction of such enzymes, their interactions and the changes they cause. It is also believed that these are responsible for the textural softening of fruit such as tomato which greatly affects their use by the processing industry and consumers (for review see Huber 1983 and Fischer and Bennett 1991).

2.1 Polygalacturonase

Polygalacturonic acid is a major component of cell walls and is often particularly abundant in fruit cells. In tomato it had long been thought that the large de novo synthesis of polygalacturonase [PG, which degrades polygalacturonic acid (pectin) to galacturonic acid monomers] at the onset of ripening was the major cause of softening (Huber 1983; Bennett and DellaPenna 1987; Brady 1987). The evidence for this included the correlation of increasing PG activity with increasing softness of the fruit (Hobson 1965) and the much reduced levels of PG in mutant tomato fruit which fail to soften (Brady et al. 1983). Although the inhibition of PG activity to as little as 0.5% of wild-type levels by a PG antisense transgene resulted in fruit more resistant to splitting (Smith et al. 1988), there was, in initial tests, little effect on softening (Schuch et al. 1991). Kramer et al. (1992) subsequently reported that PG inhibition by an antisense gene (Sheehy et al. 1988) to 1% of the control level did impair softening but these experiments used different varieties of tomato. These data suggest that although PG has some role in softening it is clearly not the only determinant. Fruit from PG antisense plants do show increased soluble polyuronide content and improved processing quality over controls (Smith et al. 1990a; Schuch et al. 1991). Kramer et al. (1992) also reported increased disease resistance to the post-harvest pathogens, *Geotrichum candidum* and *Rhizopus stolonifer*. However, there was no increased resistance to *Colletotrichum gloeosporioides* using the reduced PG plants of Smith et al. (1988; Cooper 1993). This could be because of varietal differences in the host tomatoes used or greater invasiveness of *C. gloeosporioides* compared to *G. candidum* and *R. stolonifer*.

Polygalacturonase had also been proposed to have a role in initiating fruit

gaseous hormone has long been recognised as playing a major role in the ripening of climacteric fruit such as the tomato (see Sect. 3). However, in the PG antisense plants, neither ethylene synthesis nor ripening was affected (SMITH et al. 1990a), implying that the generation of pectin fragments is not responsible for triggering the increase in ethylene synthesis.

The same part of the PG cDNA used by SMITH et al. (1988) in their antisense gene was also expressed in the sense orientation as a control but unexpectedly also inhibited PG production at least as effectively as the antisense. No biochemical differences were observed in the fruit ripening of tomatoes in which PG activity had been inhibited by the sense compared to the antisense gene (SMITH et al. 1990b).

2.2 Pectin Esterase

The galacturonosyl residues in the pectin of plant cell walls are frequently methoxylated at their C6 carboxyl groups (FRY 1988). This esterification may have a physiological role: it could prevent the formation of Ca^{2+} cross-links in pectins (FRY 1986) and decreases the susceptibility of pectin to degradation by polygalacturonase (PRESSEY and AVANTS 1982; SEYMOUR et al. 1987). It is possible that before incorporation of pectin into the cell wall, the carboxyl groups of its galacturonosyl residues are already highly methoxylated and once in the cell wall the pectin can be de-esterified to varying degrees by the action of pectin esterase (PE; KAUSS and HASSID 1967; ROBERTS 1990). PE is expressed throughout fruit development, increasing two- to threefold during ripening (HOBSON 1963). The role of methylesterification and the enzyme PE has been addressed by antisense gene-mediated suppression (TIEMAN et al. 1992; HALL et al. 1993) using a PE cDNA expressed during ripening. Both groups of authors reported the reduction to less than 10% of wild-type levels of PE activity in fruit from plants transformed with a PE antisense gene, confirming that these cloned sequences indeed encode functional PEs. Pectin remained more extensively methoxylated in the antisense fruit, confirming this enzyme's role in vivo. In both cases activity could not be reduced below 7% of normal PE activity with the antisense genes used and the expression of a different isozyme of PE encoded by a gene not homologous to any cloned PE genes was suggested as the explanation for the residual activity. TIEMAN et al. (1992) reported reduced PG activity in the fruit, corroborating earlier suggestions that PE and PG act synergistically upon the same polymer (PRESSEY and AVANTS 1982; SEYMOUR et al. 1987). Their observations that PE antisense fruit contained less EDTA-soluble polyuronides compared with wild type also provide support for the Ca^{2+} bridge theory of pectin cross-linking. The reduction of PE activity improves the processing of the tomatoes (GRIERSON and SCHUCH 1993), although HALL et al. (1993) report no reduced softening of PE antisense fruit.

2.3 Polygalacturonase and Pectin Esterase

There have been two reports of simultaneous inhibition of PG and PE with homologous suppressing transgenes. In one experiment, a transcriptional fusion of PG and PE sense sequences inhibited the expression of both genes (Seymour et al. 1993). Grierson and Schuch (1993) reported that as well as combining the improved qualities of both the single effect lines (see Sects. 2.1 and 2.2) the fruit from plants with the double construct had additional improvements in other processing parameters. Pear et al. (1993) described the successful inhibition of both these enzymes by two separate antisense genes incorporated into the host genome as a single locus, but no physiological data were described. It is likely that such combinatorial constructs will prove very useful in analysing complex biochemical pathways and enzyme interactions.

3 Ethylene

The pivotal role of the phytohormone ethylene in the ripening of climacteric fruit (e.g. tomato, apple, banana and avocado) has been extensively studied over the last few decades (Burg 1962; Yang 1981; Yang and Hoffman 1984) although humans had already unwittingly exploited the effects of ethylene for many centuries. For example, the Chinese hastened the ripening of fruit by exposing them to burning incense (Salisbury and Ross 1969). The pathway for the biosynthesis of ethylene is the same in all higher plants studied (Yang and Hoffman 1984). The two enzymes specific to the pathway are ACC-synthase and ACC-oxidase (for review see Kende 1993): ACC-synthase converts S-adenosyl methionine to aminocyclopropane-1-carboxylic acid (ACC) and ACC-oxidase converts ACC to ethylene. Antisense transgenes have been used in the investigation of both enzymes and played a particularly significant role in the identification of ACC-oxidase.

3.1 ACC-oxidase

Although genes encoding ACC-synthase had been cloned using information gained from protein purification (Sato and Theologis 1989; van der Straeten et al. 1990; Olson et al. 1991; Nakajima et al. 1990), ACC-oxidase could not be purified and no molecular probes were available. The identification of the gene encoding this enzyme in tomato was accomplished as follows: a cDNA (pTOM13) had been isolated from a tomato fruit ripening library and the expression of its homologous mRNA correlated with ethylene synthesis (Smith et al. 1986; Holdsworth et al. 1987). An antisense gene constructed from this cDNA strongly inhibited expression of ACC-oxidase activity in transgenic tomatoes (Hamilton et al. 1990). Subsequent heterologous expression of related cDNAs showed that these alone encoded ACC-oxidase (HAMILTON et al. 1991; SPANU et al. 1991). At about the same time, significant sequence homology between pTOM13 and a 2-oxoglutarate-dependent dioxygenase, flavanone 3-hydroxylase, was discovered (A. Prescott, personal communication, cited in HAMILTON et al. 1990). Using the published extraction and assay method for the flavanone-3-hydroxylase, VERVERIDIS and JOHN (1991) were able to extract ACC-oxidase activity from melon fruit. Since then considerable progress has been made in the purification of ACC-oxidase and cloning of ACC-oxidase genes from a number of plants (for review see KENDE 1993 and PRESCOTT 1993).

As well as providing evidence that pTOM13 encoded ACC-oxidase, the antisense experiment produced plants in which ethylene synthesis in fruit was reduced to around 5% of the normal level. These fruit did not acquire the normal full redness and were much more resistant to shrivelling upon prolonged storage (HAMILTON et al. 1990). A more thorough study of these plants by PICTON et al. (1993) investigated further the role of ethylene in fruit ripening. This revealed that detaching the fruit from the plant prior to ripening enhanced the subsequent delayed ripening phenotype. This may be the result of preventing transport of some unidentified factor into or out of the fruit. It has recently been shown that cytokinins are exported from tomato fruit during ripening (MARTINEAU et al. 1994). This type of plant growth regulator has been implicated as an inhibitor of ripening (ABDEL-KADER et al. 1966; DAVEY and VAN STADEN 1978) and so is a prime candidate for this hypothetical mobile factor. Based on this, normal ripening might depend upon the ratio of ethylene to cytokinins or different aspects of ripening may be affected by the two hormones. Despite the visually obvious delay in ripening, fruit from these plants showed surprisingly little difference from wild type in their expression of a variety of ripening-related genes (PICTON et al. 1993). It is possible that the genes studied might not include those critically involved in the obvious changes in ripening. Alternatively, ethylene might modify the translation of those mRNAs or the activity and stability of their protein products.

Ethylene is involved in other programmed degenerative processes such as leaf and flower senescence. The onset of leaf senescence was delayed significantly in ACC-oxidase antisense plants, confirming the suspected role of ethylene in this process (PICTON et al. 1993).

Ethylene is necessary for the full development of quality attributes associated with ripening. However, the continued production of ethylene quickly results in a product which cannot be transported or stored and is easily colonised by fungi and bacteria. Ethylene also hastens the senescence of leafy vegetables and cut flowers. The spoilage of produce as a result of these effects causes enormous losses to growers, vendors and consumers and so it is likely that the manipulation of ethylene biosynthesis with antisense (and probably sense) genes will have major agronomic application in the future.

3.2 ACC-synthase

Tomato plants with a greater reduction in ethylene synthesis were produced by Oeller et al. (1991) using an ACC-synthase antisense construct. The retardation of ripening was more pronounced in these than with the ACC-oxidase antisense gene. Oeller et al. (1991) suggest that because ACC-synthase is less stable than ACC-oxidase the former might be more susceptible to antisense inhibition. However, considerably more ACC-synthase antisense plants were produced, thus increasing the chance of identifying a severely suppressed line. Furthermore, the ACC-oxidase inhibition was by a single-copy antisense gene whereas ten copies of the ACC-synthase antisense gene at one locus were present in the line described. Using these plants it was shown that softening, colour production, aroma development and the rise in respiration were all ultimately induced by ethylene. Furthermore, these physiological changes could be restored by the addition of exogenous ethylene. Although ethylene's role as the ripening hormone has long been accepted, this was based primarily on physiological evidence (Yang and Hoffman 1984; Brady 1987). The selective inhibition of ethylene biosynthesis genes has provided complementary genetic evidence confirming the central role of ethylene in climacteric fruit ripening.

3.3 E8

Another antisense transgene which, more unexpectedly, affected ethylene synthesis was constructed from the ripening-related cDNA E8 in order to investigate the function of this gene (Peñarrubia et al. 1992). RNA homologous with the E8 clone is expressed concomitantly with ethylene synthesis in ripening fruit (Lincoln and Fischer 1988). This cDNA has 34% amino acid sequence identity with ACC-oxidase (Deikman and Fischer 1988) and so likely encodes a 2-oxoglutarate-dependent dioxygenase. Ripening fruit from E8-antisense plants which had less than 10% of normal levels of the E8 protein, synthesised six times more ethylene than did wild-type controls. The reason for this is not yet known and awaits the measurement of ACC-synthase and ACC-oxidase activity.

4 Phytoene Synthase

The second enigmatic cDNA to disclose its secrets upon inhibition with an antisense gene was another ripening-related clone, pTOM5 (Bird et al. 1991). A number of plants transformed with the pTOM5 antisense gene had greatly reduced amounts of pTOM5-homologous RNA. These plants produced yellow fruit with reduced carotenoid content and flowers with pale yellow corollas. Bramley et al. (1992) showed that these plants accumulated phytoene, the first carotenoid in the pathway. Bartley et al. (1992) demonstrated that expression of the pTOM5 cDNA complemented the carotenoidless *crtB* mutant of *Rhodobacter capsulatus*. Together these results demonstrated that the pTOM5 cDNA encodes phytoene synthase (PSY) in tomato. Like the experiments with ACC-oxidase, the antisense approach with pTOM5 not only provided important evidence as to the function of the protein encoded by a cDNA clone, but also resulted in the generation of plants in which the result of reduced enzyme activity could be studied. In these plants the synthesis of carotenoids in leaves was not affected, implying there is a different isozyme of PSY in leaves not susceptible to inhibition by this antisense gene. A fuller spectrum of the effects of disturbing carotenoid production was observed in subsequent experiments in which FRAY and GRIERSON (1993) showed that the naturally occurring tomato mutant *yellow flesh* could be complemented with a functional phytoene synthase sense transgene. Although this mutant has impaired carotenoid synthesis in ripening fruit and flowers, the plants are otherwise phenotypically normal, indicating that the phytoene synthase responsible for carotenoid production in vegetative tissues is unaffected. In the population of transgenic plants obtained, as well as individuals resembling wild-type tomatoes due to the restoration of normal lycopene production, there were plants exhibiting the striking effects of underproduction of carotenoids not only in their fruit and flowers but also in their leaves. The visual manifestation of this was photobleaching of the chlorophyll resulting in yellow leaves and pale, almost silver, fruit and flowers. Both the transgene transcript and the truncated transcript from the mutated endogenous gene were much reduced in abundance. This effect is very similar to previously recorded cases in which the transformation with additional sense copies of dihydroflavanol reductase (VAN DER KROL et al. 1990) and chalcone synthase (NAPOLI et al. 1990) in petunia, PG in tomato (SMITH et al. 1990b), and β-1,3-glucanase (DE CARVALHO et al. 1992) and chitinase (HART et al. 1992) in *Nicotiana sylvestris* caused the suppression of the expression of the endogenous gene.

Leaves do contain PSY activity but no mRNAs which hybridise with the pTOM5 cDNA in RNA gel blot analyses (R. Fray, unpublished data). However, the result of FRAY and GRIERSON (1993) suggest the pTOM5 cDNA sense sequence apparently can suppress this gene. BARTLEY et al. (1992) suggest that the pTOM5 antisense gene may have failed to suppress carotenoid production in leaves (BIRD et al. 1991) because it was synthesised from a part of the pTOM5 cDNA which lacked homology with the gene expressed in leaves. However, in a small number of transgenic plants, a pTOM5 antisense gene constructed from the entire coding sequence similarly failed to reduce carotenoid synthesis in leaves while severely impairing their production in fruit (R. Fray, unpublished data).

5 Mechanism of Action of Antisense and Sense Suppression

Numerous examples of what MATZKE has termed "homology-dependent gene silencing" exist (JORGENSEN 1992; MATZKE et al. 1994) but only some of these have been observed in experiments with fruit-ripening genes. Those described in this chapter fall into the categories of reciprocal and non-reciprocal, ectopic, *trans*-inactivation and the following discussion will be confined to these. The term "co-suppression" will be used to indicate concomitant suppression of the transgene's transcript (sense or antisense) with the target, endogenous gene's transcript. None of the experiments with fruit-ripening genes published to date have been designed specifically to address the nature of the suppression caused by antisense and sense transgenes. Instead they have been aimed at investigating the biochemical and physiological roles of proteins encoded by those genes. As a result the comparison of data for the purpose of studying suppression mechanisms is greatly complicated by differences in tomato variety used, the gene studied and the types of analyses performed. In cases where antisense transgene inheritance has been described the phenotype segregates with the suppressing transgene: plants which have lost the transgene recover a wild-type phenotype, indicating there has been no lasting effect upon the endogenous gene (SMITH et al. 1988; HAMILTON et al. 1990; TIEMAN et al. 1992; HALL et al. 1993). There are no published data concerning the segregation of a suppressed phenotype with a sense transgene in tomato. One consistent feature of antisense and sense gene suppression is the reduction of steady state levels of the target gene's mRNA. How this occurs and at which stage in the mRNA's normal biogenesis its accumulation is blocked is not clear. It has also been suggested that the mechanisms of antisense and sense gene-mediated suppression may be the same (GRIERSON et al. 1991; JORGENSEN 1992). Two broad but distinct schools of thought have emerged which unite the mechanisms of sense and antisense gene-mediated suppression: the first envisages some form of transient pairing of homologous DNA sequences. It is suggested that this would result in the alteration of the chromatin structure but not the primary sequence of either or both (in cases of co-suppression) loci (JORGENSEN 1992). The most likely effect of this would be the perturbation of some aspect of gene transcription. The second (for review see MURRAY 1992) involves antisense RNA-mediated suppression and relies on the specific base pairing of the transgene's transcript with the target gene's transcript resulting in the degradation of the latter (and possibly also the former in examples exhibiting co-suppression). This could provide a mechanism for sense gene-mediated suppression if the sense transgene gives rise to unintended antisense transcripts from promoters within or external to the T-DNA insert (GRIERSON et al. 1991). A third model which accounts only for sense gene suppression invokes the existence of regulatory circuits based on the perception of concentrations of specific mRNAs. It is proposed that these exist in discontinuous stable states lying between regions of instability and would apply to all molecules with the appropriate recognition sequence (MEINS 1989; JORGENSEN 1992). A fourth model which could include both sense and antisense-mediated suppression but which has received little attention is of an RNA-DNA interaction in which either the sense or the antisense transgene transcript base pairs with the target gene. This would most likely directly or indirectly interfere with transcription. Recently, the results of WASSENEGGER et al. (1994) have indicated the possibility that viroid RNA can suppress the expression of viroid cDNA transgenes.

One major distinction between the above models is that the first and fourth would probably affect transcription whereas the second and third would act some way downstream of this. Using nuclear run-on assays, SHEEHY et al. (1988) found there was no difference in endogenous PG gene transcription in nuclei isolated from wild-type plants and PG antisense plants. Although it is possible that the process of nuclear isolation in some way relieved the effect of the antisense gene, this evidence indicates that transcription initiation is unaffected. Similar experiments have not been carried out for examples of sense gene suppression in tomato. However, in other plants where sense transgenes have been found to inhibit the expression of homologous, endogenous genes, again no effect upon the transcription of these genes in isolated nuclei was found (MOL et al. 1991; DE CARVALHO et al. 1992). One criticism of nuclear run-on experiments is that they do not exclude a mechanism involving the disturbance of transcription elongation through the 3′ end of the gene resulting in the production of incomplete, unstable transcripts. We have obtained evidence corroborating and extending the findings of SHEEHY et al. (1988) through our analysis of the ACC-oxidase antisense plants: although the amount of full-length ACC-oxidase mRNA is reduced in wounded leaves (HAMILTON et al. 1990), fragments corresponding to the extreme 3' end of the mRNA are as abundant as in leaves from wild-type plants (A. Hamilton, unpublished data). This suggests that transcription of the endogenous gene is unaffected. We can exclude antisense-induced internal transcription initiation since we detect as much ACC-oxidase pre-mRNA in antisense plants as in wild-type plants using an intron 1-exon 2 probe in ribonuclease protection assays (A. Hamilton, unpublished data). These results suggest that in the ACC-oxidase antisense plants, the ACC-oxidase gene is transcribed fully and accurately, but the antisense RNA (presumably by base pairing) interacts with either the primary transcript or the mRNA, causing the degradation of the latter (apart from the 3' sequences). The experiments with tomato fruit-ripening genes have made less of a contribution to the understanding of sense gene suppression. In agreement with the results of VAN DER KROL et al. (1990), NAPOLI et al. (1990), DE CARVALHO et al. (1992) and HART et al. (1992), we have found that a sense transgene can be either over-expressed or co-suppressed with the endogenous gene (FRAY and GRIERSON 1993). We have not observed a dependence on homozygosity for the suppressing locus with either PG or PSY sense genes (SMITH et al. 1990b; FRAY and GRIERSON 1993).

Another interesting aspect of the antisense interaction is the occurrence of co-suppression of the antisense transcript. This has a bearing on the nature of the mechanism: if the hypothetical sense RNA-antisense RNA duplex is a substrate for degradative enzymes, one would expect that when the level of sense RNA transcription increased (due to normal switching on of its expression) the amount of antisense RNA measured would drop if it was (as is usually the case) not subject to differential transcriptional regulation. With some examples of antisense fruit-ripening genes insufficient data have been published to determine whether co-suppression is occurring (Kramer et al. 1992; Peñarrubia et al. 1992; Tieman et al. 1992; Pear et al. 1993). Co-suppression was observed with antisense genes for PG (Smith et al. 1988 and possibly Sheehy et al. 1988), ACC-oxidase (Hamilton et al. 1990) and possibly PE (Hall et al. 1993). However, we have found in the case of ACC-oxidase that this phenomenon was only observed in the original primary transformant and not in homozygous progeny or hemizygous plants produced by backcrossing homozygous ACC-oxidase antisense plants with wild types (A. Hamilton, unpublished data). Furthermore, no co-suppression was observed in ACC-synthase antisense plants (Oeller et al. 1991). This is in line with results from other antisense genes in plants in which co-suppression is not observed (Stockhaus et al. 1990). On this evidence it seems that any down-regulation of the antisense transcript is coincidental to that of the sense transcript. It is possible that the antisense RNA interacts only transiently with the target sense RNA, thereby interfering with some aspect(s) of its normal processing and transport and resulting in its destruction. The fate of the antisense transcript itself may be more variable, perhaps because it is foreign. If it is not degraded after its first interaction it may be free to interfere with many more of the target mRNAs. This may explain the success of extremely low amounts of antisense RNA in inhibiting apparently more highly expressed genes (van der Krol et al. 1988; Stockhaus et al. 1990; Bird et al. 1991).

6 Conclusions and Outlook

The effectiveness and utility of suppressing transgenes has opened a new area of continuity between molecular biology, biochemistry, physiology, plant breeding and horticulture. The technical simplicity will allow most plant biologists access to this powerful tool although our limited understanding of its mechanisms still means there is a significant element of uncertainty concerning which sequence to use and the interpretation of certain results. Nevertheless, questions have been answered and others raised concerning long-standing dogmas. New genes have been identified and characterised, allowing major gaps in our knowledge of plant biochemistry to be filled. New systems in which single or multiple known genes are severely inhibited are becoming available at an increasing rate. In our laboratory alone, eight new lines into which antisense and sense genes of previously uncharacterised ripening related cDNAs have been transformed are already being analysed. From a practical standpoint the technology has been highly successful, allowing the intelligent manipulation of a major crop species, tomato, with the production of a variety of new lines with improved agronomic performance.

References

Abdel-Kader AS, Leonard LM, Maxie EC (1966) Effect of growth regulating substances on the ripening and shelf life of tomatoes. Hort Sci 1: 90–91

Baldwin EA, Pressey R (1988) Tomato polygalacturonase elicits ethylene production in tomato fruit. J Am Soc Hort Sci 113: 92–95

Bartley GE, Viitanen PV, Bacot KO, Scolnik PA (1992) A tomato gene expressed during fruit ripening encodes an enzyme of the carotenoid biosynthesis pathway. J Biol Chem 267: 5036–5039

Bennett AB, DellaPenna D (1987) Polygalacturonase: its importance and regulation in ripening. In: Thompson WW, Nothnagel E, Huffaker R (eds) Plant senescence: its biochemistry and physiology. American Society of Plant Physiology, Rockville, MD, pp 98–107

Bird CR, Ray JA, Fletcher JD, Boniwell JM, Bird AS, Blain I, Bramley PM, Schuch W (1991) Using antisense RNA to study gene function: inhibition of carotenoid biosynthesis in transgenic tomatoes. Biotechnology 9: 635–639

Brady CJ (1987) Fruit ripening. Annu Rev Plant Physiol 38: 155–178

Brady CJ, Meldrum SK, McGlasson WB, Ali ZM (1983) Differential accumulation of the molecular forms of polygalacturonase in tomato mutants. J Food Biochem 7: 7–14

Bramley P, Teulieres C, Blain I, Bird C, Schuch W (1992) Biochemical characterisation of transgenic tomato plants in which carotenoid synthesis has been inhibited through the expression of antisense RNA to pTOM5. Plant J 2: 343–344

Burg S (1962) The physiology of ethylene formation. Annu Rev Plant Physiol 13: 265–302

Cooper W (1993) Investigations into the interactions of Colletotrichum gloeosporioides and ripening tomato fruit. Ph D thesis. University of Nottingham, UK

Davey JE, Van Staden J (1978) Endogenous cytokinins in the fruit of ripening and non-ripening tomatoes. Plant Sci Lett 11: 359–364

de Carvalho F, Gheysen G, Kushnir S, van Montagu M, Inzé D, Castresana C (1992) Suppression of β-1,3-glucanase transgene expression in homozygous plants. EMBO J 11: 2595–2602

Deikman J, Fischer RL (1988) Interaction of a DNA binding factor with the 5'- flanking region of an ethylene responsive fruit ripening gene from tomato. EMBO J 7: 3315–3320

Fischer RL, Bennett AB (1991) Role of cell wall hydrolases in fruit ripening. Annu Rev Plant Physiol Plant Mol Biol 42: 675–703

Fray RG, Grierson D (1993) Identification and genetic analysis of normal and mutant phytoene synthase genes of tomato by sequencing, complementation and co-suppression. Plant Mol Biol 22: 589–602

Fry SC (1986) Cross-linking of matrix polymers in the growing cell walls of angiosperms. Annu Rev Plant Physiol 37: 165–186

Fry SC (1988) The growing plant cell wall: chemical and metabolic analysis. Wiley, New York, pp 332

Grierson D, Fray RG, Hamilton AJ, Smith CJS, Watson C (1991) Does co-suppression of sense genes in transgenic plants involve antisense RNA. Trends Biotechnol 9: 122–123

Grierson D, Schuch W (1993) Control of ripening. Philos Trans R Soc Lond [Biol] 342: 241–250

Hall LN, Tucker GA, Smith CJS, Watson CF, Seymour GB, Bundick Y, Boniwell JM, Fletcher JD, Ray JA, Schuch W, Bird CR, Grierson D (1993) Antisense inhibition of pectinesterase gene expression in transgenic tomatoes. Plant J 3: 121–129

Hamilton AJ, Lycett GW, Grierson D (1990) Antisense gene that inhibits synthesis of the hormone ethylene in transgenic plants. Nature 346: 284–287

Hamilton AJ, Bouzayen M, Grierson D (1991) Identification of a tomato gene for the ethylene-forming enzyme by expression in yeast. Proc Natl Acad Sci USA 88: 7434–7437

Hart CM, Fischer B, Neuhaus J-M, Meins F Jr (1992) Regulated inactivation of homologous gene expression in transgenic Nicotiana sylvestris plants containing a defense-related tobacco chitinase gene. Mol Gen Genet 235: 179–188

Hobson GE (1963) Pectinesterase in normal and abnormal tomato fruit. Biochem J 86: 358–365

Hobson GE (1965) The firmness of tomato fruit in relation to polygalacturonase activity. J Hort Sci 40: 66–72

Holdsworth MJ, Bird CJ, Ray J, Schuch W, Grierson D (1987) Structure and expression of an ethylene-related mRNA from tomato. Nucleic Acids Res 15: 731–739

Huber DJ (1983) The role of cell wall hydrolases in fruit softening. Hort Rev 5: 169–219

Jorgensen R (1992) Silencing of plant genes by homologous transgenes. Agbiotech News Inf 4: 265–273

Kauss H, Hassid WZ (1967) Enzymic introduction of the methyl ester groups of pectin. J Biol Chem 242: 3449–3453

Kende H (1993) Ethylene biosynthesis. Annu Rev Plant Physiol Plant Mol Biol 44: 283–307

Kramer M, Sanders R, Bolkan H, Waters C, Sheehy RE, Hiatt WR (1992) Postharvest evaluation of transgenic tomatoes with reduced levels of polygalacturonase: processing, firmness and disease resistance. Postharvest Biol Technol 1:241–255

Lincoln JE, Fischer RL (1988) Diverse mechanisms for the regulation of ethylene inducible gene expression. Mol Gen Genet 212: 71–75

Martineau B, Houck CM, Sheehy RE, Hiatt WR (1994) Fruit specific expression of the A. Tumefaciens isopentenyl transferase gene in tomato: effects on fruit ripening and defense-related gene expression in leaves. Plant J 5: 11–19

Matzke MA, Matzke AJM, Mittelsten-Scheid O (1994) Inactivation of repeated genes: DNA-DNA interaction? In: Paszkowski J (ed) Homologous recombination in plants. Kluwer, Amsterdam (in press)

McNeil M, Darvill AG, Fry S, Albersheim P (1984) Structure and function of the primary cell walls of plants. Annu Rev Biochem 53: 625–663

Meins F Jr (1989) Habituation: heritable variation in the requirement of cultured plant cells for hormones. Annu Rev Genet 23: 395–408

Mol J, van Blokland R, Kooter J (1991) More about co-suppression. Trend Biotechnol 9: 182–183

Murray JAH (ed) (1992) Antisense RNA and DNA. Wiley Liss, New York

Nakajima N, Mori H, Yamazaki K, Imaseki H (1990) Molecular cloning and sequence of a complementary DNA encoding 1-aminocyclopropane-1-carboxylate synthase induced by tissue wounding. Plant Cell Physiol 31: 1021–1029

Napoli C, Lemieux C, Jorgensen R (1990) Introduction of a chimaeric chalcone synthase gene into petunia results in reversible co-suppression of homologous genes in trans. Plant Cell 2: 279–289

Oeller PW, Min-Wong L, Taylor LP, Pike DA, Theologis A (1991) Reversible inhibition of tomato fruit senescence by antisense RNA. Science 254: 437–439

Olson DC, White JA, Edelman L, Harkins RN, Kende H (1991) Differential expression of two tomato genes for 1-aminocyclopropane-1-carboxylate synthase in tomato fruits. Proc Natl Acad Sci USA 88: 5340–5344

Pear JR, Sanders RA, Summerfelt KR, Martineau B, Hiatt WR (1993) Simultaneous inhibition of two tomato fruit cell wall hydrolases, pectinmethylesterase and polygalacturonase with antisense gene constructs. Antisense Res Dev 3: 181–190

Peñarrubia L, Aguilar M, Margossian L, Fischer RL (1992) An antisense gene stimulates ethylene hormone production during tomato fruit ripening. Plant Cell 4: 681–687

Picton S, Barton SL, Bouzayen M, Hamilton AJ, Grierson D (1993) Altered fruit ripening and leaf senescence in tomatoes expressing an antisense ethylene forming-enzyme transgene. Plant J 3: 469–481

Prescott AG (1993) A dilemma of dioxygenases (or where biochemistry and molecular biology fail to meet). J Exp Bot 44: 849–861

Pressey R, Avants JK (1982) Solubilisation of cell walls by tomato polygalacturonases: effects of pectinesterases. J Food Biochem 6: 57–74

Roberts K (1990) Structures at the plant cell surface. Curr Opin Cell Biol 2: 920–928

Salisbury FB, Ross C (1969) Plant physiology. Wadsworth, Belmont, California, pp 339

Sato T, Theologis A (1989) Cloning the mRNA encoding 1-aminocyclopropane-1-carboxylate synthase, the key enzyme for ethylene biosynthesis in plants. Proc Natl Acad Sci USA 86: 6621–6625

Schuch W, Kanczler J, Robertson D, Hobson G, Tucker G, Grierson D, Bright S, Bird C (1991) Fruit ripening characteristics of transgenic tomato fruit with altered polygalacturonase activity. Hort Sci 26: 1517–1520

Seymour GB, Lasslett Y, Tucker GA (1987) Differential effects of pectolytic enzymes on tomato polyuronides in vivo and in vitro. Phytochemistry 26: 3137–3139

Seymour GB, Fray RG, Hill P, Tucker GA (1993) Down regulation of two nonhomologous endogenous tomato genes with a single chimeric sense gene construct. Plant Mol Biol 23: 1–9

Sheehy RE, Kramer M, Hiatt WR (1988) Reduction of polygalacturonase activity in tomato fruit by antisense RNA. Proc Natl Acad Sci USA 85: 8805–8809

Smith CJS, Slater A, Grierson D (1986) Rapid appearance of an mRNA correlated with ethylene synthesis encoding a protein of molecular weight 35 000. Planta 168: 94–100

Smith CJS, Watson CF, Ray J, Bird CR, Morris PC, Schuch W, Grierson D (1988) Antisense RNA inhibition of polygalacturonase gene expression in transgenic tomatoes. Nature 334: 724–726

Smith CJS, Watson CF, Morris PC, Bird CR, Seymour GB, Gray JE, Arnold C, Tucker GA, Schuch W, Harding S, Grierson D (1990a) Inheritance and effect on ripening of antisense polygalacturonase genes in transgenic tomatoes. Plant Mol Biol 14: 369–379

Smith CJS, Watson CF, Bird CR, Ray J, Schuch W, Grierson D (1990b) Expression of a truncated tomato polygalacturonase gene inhibits expression of the endogenous gene in transgenic plants. Mol Gen Genet 224: 477–481

Spanu P, Reinhardt D, Boller T (1991) Analysis and cloning of the ethylene forming enzyme from tomato by functional expression of its mRNA in Xenopus laevis oocytes. EMBO J 10: 2007–2013

Stockhaus J, Höfer M, Renger G, Westhoff P, Wydrzynski T, Willmitzer L (1990) Anti-sense RNA efficiently inhibits formation of the 10 kd polypeptide of the photosystem II in transgenic plants: analysis of the role of the 10 kd protein. EMBO J 9: 3013–3021

Tieman DM, Harriman RW, Ramamohan G, Handa AK (1992) An antisense pectin methylesterase gene alters pectin chemistry and soluble solids in tomato fruit. Plant Cell 4: 667–679

Tigchelaar EC, McGlasson WB, Buescher RW (1978) Genetic regulation of tomato fruit ripening. J Hort Sci 17: 508–513

Van der Krol AR, Lenting PE, Veenstra J, van der Meer IM, Koes RE, Gerats AG, Mol JNM, Stuitje AR (1988) An anti-sense chalcone synthase gene in transgenic plants inhibits flower pigmentation. Nature 333: 866–869

Van der Krol AR, Mur LA, Beld M, Mol JNM, Stuitje AR (1990) Flavonoid genes in petunia: addition of a limited number of gene copies may lead to suppression of gene expression. Plant Cell 2: 291–299

Van der Straeten D, Van Wiemeersch L. Goodman HM, Van Montagu M (1990) Cloning and sequence of two different cDNAs encoding 1-aminocyclopropane-1- carboxylate synthase in tomato. Proc Natl Acad Sci USA 87: 4859–4863

Varner JE, Lin L-S (1989) Plant cell wall architecture. Cell 56: 231–239

Ververidis P, John P (1991) Complete recovery in vitro of ethylene-forming enzyme activity. Phytochemistry 30: 725–727

Wassenegger M, Heimes S, Riedel L, Sanger HL (1994) RNA-directed de novo methylation of genomic sequences in plants. Cell 76: 567–576

Yang SF (1981) Biosynthesis of ethylene and its regulation. In: Friend J, Rhodes M J C (eds) Recent advances in the biochemistry of fruit and vegetables. Academic, London, pp 89–106

Yang SF, Hoffman NE (1984) Ethylene biosynthesis and its regulation in higher plants. Annu Rev Plant Physiol 35: 155–189

Co-suppression of β-1,3-Glucanase Genes in *Nicotiana tabacum*

F. de Carvalho Niebel[1], P. Frendo[1], D. Inzé[2], M. Cornelissen[3], and M. Van Montagu[1]

1 Introduction

Recent studies have demonstrated that introduction of sense gene constructs in plants can result in transgenic plants which express the transgene but also in plants with suppressed levels of transgene expression (reviewed by Jorgensen 1991; Kooter and Mol 1993; Matzke and Matzke 1993). In many silencing cases, the transgene and homologous host genes are coordinately suppressed in the plants, a phenomenon called co-suppression (Napoli et al. 1990). Although the underlying mechanism(s) are unknown, co-suppression has been reported in different plant species, and thus may represent a new approach to the manipulation of gene expression in plants. The finding that the presence of transgenes can influence the expression of resident genes via a homology-based mechanism opens up the possibility that this type of gene regulation may be naturally

[1] Laboratorium voor Genetica, Universiteit Gent, K.L. Ledeganckstraat 35, 9000 Gent, Belgium
[2] Laboratoire Associé de l'Institut National de la Recherche Agronomique (France), Universiteit Gent, K.L. Ledeganckstraat 35, 9000 Gent, Belgium
[3] Plant Genetic Systems N.V., J. Plateaustraat 22, 9000 Gent, Belgium

occurring in plants. A major challenge now is to unravel the molecular mechanism(s) responsible for transgene-induced gene silencing in plants.

This manuscript summarizes our current understanding regarding the "sense" co-suppression of β-1,3-glucanase genes in transgenic tobacco plants. We will first briefly review some relevant features of plant β-1,3-glucanases, then gene silencing results will be highlighted and discussed in relation to other silencing cases. In addition, a working model will be presented to explain the β-1,3-glucanase silencing mechanism.

1.1 β-1,3-Glucanases in Plants

β-1,3-Glucanases are hydrolytic enzymes which are induced in plants in response to various stress situations such as pathogen infection or chemical or hormonal treatments (review by Cutt and Klessig 1992; Meins et al. 1992). These observations together with the fact that β-1,3-glucanases show antifungal activities in vitro (Mauch et al. 1988; Sela-Buurlage et al. 1993) led to the assumption that they are involved in plant defence reaction mechanisms. As a consequence, β-1,3-glucanases have been studied in many plant species, resulting in the isolation and characterization of several genes in both monocot and dicot plants (reviewed by Bol et al. 1990). In addition to their possible role in defence, β-1,3-glucanases may be involved in plant developmental processes as well. For example, expression of β-1,3-glucanase genes has been associated with flowering (Neale et al. 1990; Ori et al. 1990), seed germination (Høj et al. 1989; Cordero et al. 1994; Vögeli-Lange et al. 1994) and seedling development (Wong and MacLachlan 1980). Some β-1,3-glucanase genes are also constitutively expressed in floral and root tissues of healthy plants (Lotan et al. 1989; Castresana et al. 1990; Neale et al. 1990; Coté et al. 1991).

In tobacco, β-1,3-glucanases are classified as acidic or basic isoforms on the basis of their measured or deduced isoelectric points. Even though both acidic and basic β-1,3-glucanases are induced in plants upon stress, their regulation, subcellular locations and tissue distribution are different.

Based on amino acid sequence identity, tobacco β-1,3-glucanases were classified into at least three main groups (Meins et al. 1992). The class I enzymes are basic isoforms localized in the cell vacuoles. Class II includes neutral and basic members, the extracellularly located acidic isoforms, and the non-stress-related acidic stylar β-1,3-glucanases. Class III, by contrast, represents only one acidic enzyme. Among the different members of class I and class II, sequence identities are more than 90% and 80%, respectively, while sequence identity between the three classes is only around 50%.

Vacuole targeting of class I basic β-1,3-glucanases is due to the presence of a C-terminal extension protein sequence which is absent in the extracellularly located class II acidic isoforms. The presence or absence of C-terminal extension sequences correlates with basic or acidic β-1,3-glucanase isoforms, respectively, but only for the stress-induced β-1,3-glucanases, since the non-stress-related class II acidic stylar proteins also contain a C-terminal extension sequence.

Further investigation of the localization of the remaining class II enzymes and the class III protein will help to elucidate whether vacuolar/extracellular locations of β-1,3-glucanases are correlated with different structural classes or with enzymes having basic or acidic pI values. These analyses may increase the understanding of the functional relationship between the different enzymes.

1.2 Characterization of *Nicotiana plumbaginifolia* β-1,3-Glucanase Genes

We have previously described the isolation and characterization of two genes (*gn1* and *gn2*) coding for β-1,3-glucanases in *Nicotiana plumbaginifolia* plants (DE LOOSE et al. 1988; CASTRESANA et al. 1990; GHEYSEN et al. 1990). Sequence analysis revealed that the *gn2* gene is highly homologous (98%) to class I basic vacuolar β-1,3-glucanase from *N. tabacum*. The *gn1* β-1,3-glucanase is a basic, vacuolar-located protein which constitutes a new isoform, different from the previously characterized *N. tabacum* proteins. The *gn1* β-1,3-glucanase shows approximately 75%, 50% and 54% sequence identity at the amino acid level, to the class I, II and III tobacco β-1,3-glucanases, respectively. At the DNA level, the sequence identity is about 81%, 62% and 64%.

The expression characteristics of the *gn1* gene were examined both in *N. plumbaginifolia* and in transgenic tobacco plants containing a *gn1* promoter-β-glucuronidase (GUS) gene fusion. These studies revealed that *gn1* is expressed at higher levels in the lower parts than in the upper parts of healthy plants, as tobacco class I genes, and its expression is highly induced in leaves at local sites of bacterial infection. Importantly, both tissue-specific and stress-induced expression of the *gn1* gene is primarily controlled at the transcriptional level (CASTRESANA et al. 1990).

1.3 Sense Inhibition of β-1,3-Glucanase Genes in Transgenic Tobacco

To test whether *gn1* β-1,3-glucanase could confer increased resistance to plants against pathogen attack, transgenic plants were generated which contain a chimeric *gn1* gene, designed to produce constitutively high levels of the enzyme. This study revealed an unexpected phenomenon of transgene-induced gene silencing, where both the transgene and homologous endogenous genes were coordinately silenced. The existence of transgenic lines with repressed or non-repressed *gn1* β-1,3-glucanase levels provided a useful experimental system for characterization of this silencing phenomenon in detail. Our analyses revealed that a post-transcriptional mechanism operates specifically on homologous genes, leading to reduced steady-state mRNA levels in the suppressed plants. Furthermore, this phenomenon depends on developmental factors and on the transgene dose. These findings suggest that silencing may be a naturally occurring pathway of gene regulation in plants.

2 Results

2.1 Analysis of *gn1* Expression in Transgenic Tobacco Plants

A chimeric gene was constructed by fusing the CaMV 35S promoter to a (P35S) 3.3-kb DNA fragment containing the entire coding region and 1.35-kb 3'-flanking sequences of the *gn1* gene. The P35S-*gn1* gene was inserted into a binary vector that also contained a chimeric neomycin phosphotransferase (*nptII*) gene, as selection marker. The resulting chimeric construct was introduced into *N. tabacum* cv. SR1 (DE CARVALHO et al. 1992) and the transgenic plants were selected on medium containing kanamycin (100 mg/l). Five primary transformants (R0) that expressed the P35S-*gn1* and contained one T-DNA locus were self-fertilized for further analysis. The progeny derived from the highest expressing R0 plant T17, which contained one locus with a complex T-DNA structure with ≈three transgenes, displayed an unexpected pattern of *gn1* gene expression. T17 plants homozygous for the T-DNA locus had almost undetectable levels of *gn1* mRNA in all assayed tissues, while hemizygous plants showed *gn1* mRNA levels similar to those of the parental R0 T17 plant. *gn1* silencing in homozygous T17 plants was stable since the absence of *gn1* steady-state mRNA was observed in the R1, R2, R3 and R4 generations, obtained by self-pollination. In addition, crosses between R2 and R3 homozygous T17 plants with wild-type tobacco plants produced hemizygous plants with restored levels of *gn1* expression. By contrast, the chimeric *nptII* gene present on the same T-DNA locus showed an expected gene dose-response since the *nptII* mRNA accumulated to ≈twofold higher levels in homozygous than in hemizygous T17 plants. Therefore, silencing in T17 plants is specific for the P35S-*gn1* gene and is strictly correlated with the homozygous state of the transgene locus (DE CARVALHO et al. 1992).

2.2 Co-suppression of Endogenous β-1,3-Glucanase Genes

Some of the first results to show that sense transgenes can become silenced in plants were reported by NAPOLI et al. (1990) and VAN DER KROL et al. (1990). In these cases, chalcone synthase (*chs*) transgenes and homologous host *chs* genes were simultaneously silenced, a phenomenon referred to as co-suppression. Based on their observations, we decided to examine whether endogenous β-1,3-glucanase genes were also affected by a "co-suppression"-like mechanism in homozygous T17 plants. Preliminary results did not show large differences in the β-1,3-glucanase activity levels between young leaves of wild-type and suppressed T17 plants, probably because of the low enzyme quantities. Therefore, we looked for possible differences in β-1,3-glucanase levels in stressed plants, where endogenous β-1,3-glucanase levels are enhanced. The GN1 β-1,3-glucanase (pI≈9.6) was only found in extracts from hemizygous T17 plants. Furthermore, the endogenous basic β-1,3-glucanase activity detected in both T17 hemizygous and wild-type plant extracts was absent in homozygous plant extracts. Instead, the endogenous acidic β-1,3-glucanase activity was similar in all plant extracts. Taken together, these data imply that *gn1* and endogenous basic β-1,3-glucanase genes are co-suppressed in homozygous T17 plants.

The absence of endogenous basic β-1,3-glucanase activity in T17 homozygous plants correlated with strongly reduced levels of tobacco class I basic β-1,3-glucanase gene transcripts. Thus, co-suppression acts on members of the class I tobacco β-1,3-glucanase genes, but not on stress-induced acidic β-1,3-glucanase genes from other classes. This selective suppression may occur as a consequence of the higher degree of sequence identity that exists between *gn1* and class I basic isoform genes (≈81%) than between *gn1* and genes encoding the acidic class II and class III β-1,3-glucanases (≈63%). However, homology alone is not sufficient to trigger the suppression process since the *gn1* and endogenous class I genes are only silenced in homozygous T17 plants. This implies that the mechanism is induced in a transgene dose-dependent manner or that it requires interactions between the allelic homologous transgene copies.

2.3 Transgene Dose-Dependent Silencing of β-1,3-Glucanase Genes

As the inactivation of the β-1,3-glucanase genes occurs exclusively in T17-homozygous plants, silencing may thus depend on interactions between allelic *gn1* copies. To address this point, haploid tobacco plants that contained one T-DNA locus per chromosome set were generated from both hemizygous and homozygous T17 plants (DE CARVALHO et al. 1992). The expression of the P35S-*gn1* as well as of the endogenous β-1,3-glucanase gene appeared silenced in haploid tobacco plants, regardless of whether they were derived from homozygous or hemizygous T17 plants. This means that β-1,3-glucanase gene silencing does not require allelic interactions between *gn1* gene copies. The data suggest that the ratio between the transgene dose and either the genome copy or the number of endogenous β-1,3-glucanase genes may have importance. In other words, silencing might be activated in a dose-dependent way as a function of the number of transgene copies or the level of transgene expression.

2.4 Developmental Control of β-1,3-Glucanase Gene Silencing

To evaluate whether the suppression phenomenon was stably maintained during plant development, the expression of the P35S-*gn1* gene was examined in leaf tissues of homozygous and hemizygous T17 plants at different times after seed germination on soil. At all the developmental stages, levels of *gn1* mRNA were similar in hemizygous T17 plants. In contrast, in homozygous T17 plants *gn1* mRNA was abundant during the first weeks of plant development but disappeared after 7–8 weeks of development. Importantly, the expression of the endogenous class I genes ceased together with the *gn1* expression during development.

The timing of gene silencing appears to be a function of environmental conditions. Homozygous T17 plants grown in higher light intensities undergo silencing at an earlier time point. However, as plants also grow faster in this case, the suppressed plants are at a similar growth stage as plants grown in less light. This observation suggests that silencing operates at a certain stage of plant development. Alternatively, environmental conditions such as light intensity could be directly responsible for this early activation of gene silencing.

Once suppression is settled, it is maintained during later stages of plant development. In addition, analysis of several plant parts (upper, middle, and lower leaves; root and floral tissues) of suppressed homozygous T17 plants revealed that *gn1* suppression occurs in all assayed tissues.

2.5 Post-transcriptional Control of Gene Silencing

Co-suppression of β-1,3-glucanase genes in homozygous T17 plants results in highly reduced levels of both steady-state mRNA and protein. We investigated whether the mRNA levels in hemizygous and suppressed homozygous T17 plants correlate with the transcriptional activities of the β-1,3-glucanase genes. We could show that the *gn1* and the endogenous class I genes were transcribed with equal efficiencies in both homozygous and hemizygous T17 plants. This result indicates that reduced RNA accumulation in silenced homozygous plants is due to a post-transcriptional mechanism specifically affecting the *gn1* and endogenous basic class I β-1,3-glucanase transcripts.

It has previously been proposed that unexpected production of antisense RNA from a converging promoter, either within or near the inserted T-DNA, would be responsible for the downregulation of sense RNA production by the transgenes (Grierson et al. 1991). If this were the case for β-1,3-glucanase gene silencing, an external plant promoter should transcribe the P*nos-nptII* chimeric gene present downstream of the P35S-*gn1* gene in antisense orientation, unless internal promoter sequences located in the P*nos* promoter were to generate antisense *gn1* RNA molecules. However, we could not detect antisense *gn1* transcription in the nuclei of suppressed homozygous T17 plants, suggesting that an antisense-like mechanism is not involved in this silencing event.

Furthermore, analyses of nuclear RNA revealed that mature and polyadenylated *gn1* mRNA is produced in the nucleus of homozygous plants, suggesting that the post-transcriptional control of gene silencing operates mainly at RNA transport or RNA stability levels.

2.6 β-1,3-Glucanase Gene Silencing Is Not Restricted to T17

To test whether *gn1* silencing could be reproduced in plants other than T17, additional transformants containing the P35S-*gn1* gene were generated. The same T-DNA was used as before, except that the *nptII* marker gene was replaced by the hygromycin resistance (*hpt*) gene. Among 12 independent R0 transformants examined, plants S1 and S11, which contained multiple T-DNA copies, showed suppression of the *gn1* and the endogenous β-1,3-glucanase genes. This result demonstrates that β-1,3-glucanase gene silencing is a reproducible phenomenon that is not restricted to T17. Furthermore, the fact that co-suppression is seen in primary transformants shows that interaction between allelic copies is not required. However, on the basis of the lines analysed, we cannot establish a relation between copy number and suppression.

To evaluate whether suppression could be achieved by combining different loci derived from highly expressing plants, the two R0 plants S8 and S9 which carry two and three T-DNA loci, respectively, were crossed with homozygous T17 plants. In the hygromycin-resistant progeny, all plants should contain the T17 locus in a hemizygous state together with one to several of the loci provided by either the S8 or S9 parents. The T17 hemizygous locus should show a normal *gn1* mRNA level unless the interaction with loci from S8 or S9 plants were to cause silencing.

Five out of 11 plants examined from the cross between S9 and homozygous T17 showed the silenced phenotype. In these plants *gn1* silencing correlated with the presence of the S9.1 locus, which contains two transgenes. The remaining six plants expressing *gn1* carried either the S9.2 locus (one transgene), the S9.3 locus (one transgene) or both the S9.2 and S9.3 loci. Importantly, suppressed plants containing the T17 and the S9.1 locus and non-suppressed plants containing the T17 and the S9.2 and S9.3 loci had a similar number of transgenes. This implies that *gn1* suppression is activated by the combination of certain loci, i.e. T17 and S9.1, rather than by the absolute number of transgene copies.

In the progeny of the cross between S8 and homozygous T17, the presence of the S8.1 locus, which contains two transgenes, correlated in seven out of 13 cases with a silenced phenotype. However, one plant carrying the S8.1 locus displayed a just slightly reduced *gn1* mRNA level. Plants containing only the S8.2 locus (one transgene) showed either a high (3 out of 13) or a suppressed (2 out of 13) level of *gn1* mRNA. Thus in this case, no strong correlation can be established between silencing and the presence of specific loci. This may suggest that physiological conditions may affect the commitment of a locus to silence. The silencing competence may be determined by the position of integration in the plant genome or the spatial localization in the nucleus.

3 Discussion

3.1 Transgene Requirement

We have shown that introduction of a chimeric *N. plumbaginifolia* β-1,3-glucanase gene (*gn1*) in *N. tabacum* can result in silencing of both the transgene and endogenous homologous β-1,3-glucanase genes. This phenomenon has been extensively characterized in the T17 plant line, where co-suppression of β-1,3-

glucanase genes is strictly associated with plants homozygous for the T-DNA locus. However, by studying haploid tobacco T17 plants we found that this silencing phenomenon does not require allelic interactions between T-DNA copies, but is rather correlated with the transgene dose in the plant genome. In this context it should be noted that co-suppression of chitinase genes was only described in plants homozygous for the transgene locus (HART et al. 1992). In other silencing events, homozygosity at the suppressed loci has been correlated with enhanced suppression effects (ELKIND et al. 1990; ANGENENT et al. 1993; BRUSSLAN et al. 1993; MATZKE et al. 1993). Thus, some of these cases may be due to gene dose effects, as is also shown in our work.

In a second series of transformants, β-1,3-glucanase gene silencing was observed in two new primary transformants (S1 and S11) as well as in the progeny of S8 and S9 plants that were crossed with homozygous T17 plants. These results revealed that this suppression phenomenon was not restricted to the T17 line. Furthermore, the results confirmed that the homozygous state of the transgene is not an absolute requirement for silencing, since it also can be obtained in plants containing non-allelic T-DNA loci. Thus, suppression in homozygous T17 plants is likely to occur as a consequence of having increased amounts of transgene copies or levels of transgene expression relative to hemizygous plants. However, increased transgene copy number alone is not sufficient to activate silencing since the S4 primary transformant with a similar T-DNA copy number to S1 and S11 suppressed plants expresses the *gn1* gene. These observations indicate the existence of loci with different ability to promote silencing.

It appeared that the T17 locus in a hemizygous state can mediate gene silencing only when combined with specific S9 loci, independently of the T-DNA copy number. Similarly, HOBBS et al. (1993) have reported that silencing of a chimeric *gus* gene in tobacco does not correlate with increased copy number. Instead, the authors demonstrated that the T-DNA configuration of specific loci containing two linked copies was associated with reduced gene activity. However, transgene configuration is clearly not an absolute requirement since in a few cases suppression has been obtained with plants containing single-copy transgenes (VAN DER KROL et al. 1990; SEYMOUR et al. 1993). In this respect, the transgenic plants showing β-1,3-glucanase gene silencing always contained at least one locus with multiple transgene copies. This suggests that transgene configuration may be important for β-1,3-glucanase gene suppression.

Alternatively, the positions of the T17 and different S9 loci relative to each other in the nucleus may explain the variability in silencing, as proposed by JORGENSEN (1991). In this view, the transgene loci able to generate suppression, such as T17, S1, S11 and S9.1, as well as the endogenous homologous β-1,3-glucanase genes, should be localized in compatible nuclear domains. This would imply that plants containing the T17 locus and the S9.2 and S9.3 loci are not suppressed because these S9 loci are located in nuclear domains incompatible with the T17 and endogenous genes. However, this model cannot directly explain why the S9.2 and S9.3 loci are silenced when the T17 and the S9.1 loci are also present in the plants, unless the initiation of silencing is triggered in limited areas of the nucleus and then the mechanism itself affects other homologous gene sequences at other nuclear positions.

Another possibility is that the level or constitutive character of expression of individual loci is a major determinant of silencing. In such a scenario, the transgene configuration as well as the position of T-DNA integration in the genome would be indirectly involved in gene silencing by dictating the uniformity and amount of gene activity for each loci. Then, the dose-dependent β-1,3-glucanase suppression would be activated as a consequence of a (combined) high level of expression produced by specific transgene loci in the plants. This assumption favours the biochemical switch model proposed to explain silencing in plants. This model invokes a mechanism that is sensitive to a certain threshold level of gene expression (JORGENSEN 1991). According to this model, the S9.1 locus should express a higher level of *gn1* than the S9.2 and S9.3 loci, since only the S9.1 locus, and not the S9.2 and S9.3 loci, leads to suppression when combined with the T17 locus. This hypothesis could be tested by comparing the *gn1* expression levels of individual S9.1, S9.2 and S9.3 loci.

3.2 Homology Requirement

At present it seems that there are at least two different silencing mechanisms in plants, regulated at the transcriptional or post-transcriptional level. Transcriptional silencing has been reported for homologous *A1* transgenes (MEYER et al. 1993) as well as for chimeric *cab* transgenes and endogenous genes having sequence homology restricted to the promoter and 14 nucleotides of the leader (BRUSSLAN et al. 1993). In contrast, post-transcriptional silencing has been reported for genes where the regions of homology are at the coding region (MOL et al. 1991; LINDBO et al. 1993; this work). Analysis of an additional number of silencing cases is necessary to elucidate whether a correlation exists between the type of silencing control (transcriptional or post-transcriptional) and the location of the homologous regions on the suppressed genes (promoter or coding region homology).

The suppression of β-1,3-glucanase genes in tobacco is directed specifically to class I β-1,3-glucanase genes, members of the same group of basic isoforms, located in the plant cell vacuoles. This selective suppression is most likely due to the higher degree of sequence identity that exists between *gn1* and class I genes (≈82%) than between *gn1* and class II and III members (≈63%). Similarly, Angenent et al. (1993, 1994) reported that the petunia homeotic genes *fbp1* and *fbp2* can only suppress their homologs and other very homologous genes, while members with about 30% sequence divergence are not affected.

We looked for sequence homology common to the coding regions of *gn1* and basic β-1,3-glucanase which was absent in the coding region of acidic β-1,3-glucanases. On average, regions of mismatching between them are not larger than 9 bp, except for the 3' part of the coding region, which shows a divergence of ≈50–70 bp. Thus the specificity of this co-suppression should require sequence information present in short sequences common to the transcripts of the *gn1* and class I basic β-1,3-glucanase genes and absent in the acidic counterparts. In this respect, it would be interesting to perform a genetic deletion analysis to identify the region(s) important for class I β-1,3-glucanase gene silencing.

Several reports have demonstrated that the introduction of truncated genes in plants can also lead to silencing, even though the corresponding proteins are not produced. These results, obtained with plant, bacterial or viral genes (Smith et al. 1990; Goring et al. 1991; Lindbo et al. 1993), suggest that in general the transgene proteins are not important for the generation of silencing. In contrast, Elkind et al. (1990) proposed that the protein produced by a *pal* transgene may be involved in the induction of silencing of the endogenous *pal* gene in tobacco. It is possible that in some cases the protein produced by the transgene would generate the signals required to activate gene silencing, and a common silencing mechanism would then operate for different genes. At present we have no evidence suggesting that the *gn1* β-1,3-glucanase protein is important for gene silencing. The analysis of plants encoding truncated GN1 could address this point.

3.3 β-1,3-Glucanase Gene Silencing: Possible Mechanisms

Several parameters were identified that are important for co-suppression of β-1,3-glucanase genes in tobacco. Silencing depends on the transgene dosage, but only of specific loci which are competent to promote silencing. We believe that this competence to silence depends mainly on the levels and continuity of expression of specific loci. However, T-DNA configuration and position of integration may also influence this competence. Furthermore, silencing is erased during reproduction and is again established at a certain stage of development. It is still not known which are the developmental signals that switch on the silencing mechanism, but once silencing is settled, it is maintained. This suggests that maintenance of silencing requires a constant process, signal or imprint. Since β-1,3-glucanase silencing appears to be a post-transcriptionally regulated process, maintenance of silencing could rely on the constant transcription rate of the affected genes in the cell nuclei. For all models it will be crucial to explain why spliced, polyadenylated *gn1* mRNA accumulates in the nucleus of homozygous T17 plants. This observation suggests that the absence of a significant cytoplasmic steady-state level of *gn1* mRNA in the suppressed plant cells is due to either a block of *gn1* mRNA transport from the nucleus to the cytoplasm or an increased turnover of transported transcripts.

Transport of mRNA from the nucleus to the cytoplasm involves RNA-protein complexes. Furthermore, transport of different classes of RNA is mediated by specific protein factors. These processes are energy requiring and saturable by increased concentrations of different RNAs (Dargemont and Kuhn 1992; Jarmolowski et al. 1994). It can be envisaged that *gn1* mRNA accumulating in the nucleus may be the triggering event that blocks its own transport. Thus, a protein factor that is essential for export of the class I β-1,3-glucanase mRNAs may be saturable, or become titrated, resulting in strongly reduced transport of these transcripts. RNA binding proteins can show different binding affinities to different

RNA sequences (Burd and Dreyfuss 1994). The developmental control of gene silencing could then be a mere consequence of a permanent shortage of this transport protein. Such a shortage could result from decreased production of the transport protein or from the appearance of an mRNA with higher affinity to it. The constant level of *gn1* transcript production would, under such conditions, result in a maintenance of silencing. Alternatively, a similar dose-dependent control could be operating at the cytoplasmic level, but in this case the saturable protein factor would have a stabilization rather than a transport function.

It is possible that the β-1,3-glucanase suppression also operates in a way similar to that proposed by Lindbo et al. (1993). These authors found that a viral coat protein transgene RNA is suppressed in transgenic tobacco upon viral infection. They proposed that the signal for activation of silencing is cytoplasmic since the virus is located in the cytoplasm. Thus, a protein factor would bind to the RNA molecules, rendering them unstable. It still needs to be determined which protein is involved and which signal initiates recognition of the RNA only at a certain moment.

It should be stressed that the models proposed above aim to explain the specific suppression of homologous *gn1* and endogenous β-1,3-glucanase genes. Similar control mechanisms could be envisaged for other post-transcriptional cases of gene silencing. However, one has to assume that for each gene, specific interactions with protein factor(s) would be required for silencing to occur. Such precise control could explain the high degree of specificity that exists for each silencing event, since only specific homologous transgenes and endogenous genes are suppressed. Still, the fact that gene silencing can occur with a variety of genes in different plant systems favours a general process.

The generality of silencing would have to rely on a still unknown gene regulation control that can sense the superfluous RNA produced by the silencing-affected genes, leading to specific RNA destruction. Future experiments should aim at clarifying whether silencing requires other factors in addition to nucleic acids.

Acknowledgements. The authors wish to thank Andreas Niebel for critical reading of the manuscript. This work was supported by grants from the Belgian Programme on Interuniversity Poles of Attraction (Prime Minister's Office, Science Policy Programming, #38) and the "Vlaamse Actieprogramma Biotechnologie" (ETC 002). F.d.C.N. is indebted to the Conselho Nacional de Desenvolvimento Científico e Technológico (CNPq 200353/90-0; Brazil) for a predoctoral fellowship; P.F. was supported by an EMBO long-term fellowship; D.I. is a Research Director of the Institut National de la Recherche Agronomique (France).

References

Angenent GC, Franken J, Busscher M, Colombo L, van Tunen AJ (1993) Petal and stamen formation in petunia is regulated by the homeotic gene fbpl. Plant J 4: 101–112

Angenent GC, Franken J, Busscher M, Weiss D, van Tunen AJ (1994) Co-suppression of the petunia homeotic gene fbp2 affects the identity of the generative meristem. Plant J 5: 33–34

Bol JF, Linthorst HJM, Cornelissen BJC (1990) Plant pathogenesis-related proteins induced by virus infection. Annu Rev Phytopathol 28: 113–138

Brusslan JA, Karlin-Neumann GA, Huang L, Tobin EM (1993) An Arabidopsis mutant with a reduced level of cab140 RNA is a result of cosuppression. Plant Cell 5: 667–677

Burd CG, Dreyfuss G (1994) RNA binding specificity of hnRNP A1: significance of hnRNP A1 high-affinity binding sites in pre-mRNA splicing. EMBO J 13: 1197–1204

Castresana C, de Carvalho F, Gheysen G, Habets M, Inzé D, Van Montagu M (1990) Tissue-specific and pathogen-induced regulation of a Nicotiana plumbaginifolia β-1,3-glucanase gene. Plant Cell 2: 1131–1143

Cordero MJ, Raventós D, San Segundo B (1994) Differential expression and induction of chitinases and β-1,3-glucanases in response to fungal infection during germination of maize seeds. Mol Plant Microbe Interact 7: 23–31

Coté F, Cutt JR, Asselin A, Klessig DF (1991) Pathogenesis-related acidic β-1,3-glucanase genes of tobacco are regulated by both stress and developmental signals. Mol Plant Microbe Interact 4: 173–181

Cutt JR, Klessig DF (1992) Pathogenesis-related proteins. In: Boller T, Meins F (eds) Genes involved in plant defense. Springer, Vienna New York, pp 209–243 (Plant gene research series)

Dargemont C, Kühn LC (1992) Export of mRNA from microinjected nuclei of Xenopus laevis oocytes. J Cell Biol 118: 1–9

de Carvalho F, Gheysen G, Kushnir S, Van Montagu M, Inzé D, Castresana C (1992) Suppression of β-1,3-glucanase transgene expression in homozygous plants. EMBO J 11: 2595–2602

De Loose M, Alliotte T, Gheysen G, Genetello C, Gielen J, Soetaert P, Van Montagu M, Inzé D (1988) Primary structure of a hormonally regulated β-glucanase of Nicotiana plumbaginifolia. Gene 70: 13–23

Elkind Y, Edwards R, Mavandad M, Hedrick SA, Ribak O, Dixon RA, Lamb CJ (1990) Abnormal plant development and down-regulation of phenylpropanoid biosynthesis in transgenic tobacco containing a heterologous phenylalanine ammonia-lyase gene. Proc Natl Acad Sci USA 87: 9057–9061

Gheysen G, Inzé D, Soetaert P, Van Montagu M, Castresana C (1990) Sequence of a Nicotiana plumbaginifolia β(1,3)-glucanase gene encoding a vacuolar isoform. Nucleic Acids Res 18: 6685

Goring DR, Thomson L, Rothstein SJ (1991) Transformation of a partial nopaline synthase gene into tobacco suppresses the expression of a resident wild-type gene. Proc Natl Acad Sci USA 88: 1770–1774

Grierson D, Fray RG, Hamilton AJ, Smith CJS, Watson CF (1991) Does co-suppression of sense genes in transgenic plants involve antisense RNA. Trends Biotechnol 9: 122–123

Hart CM, Fischer B, Neuhaus J-M, Meins F Jr (1992) Regulated inactivation of homologous gene expression in transgenic Nicotiana sylvestris plants containing a defense-related tobacco chitinase gene. Mol Gen Genet 235: 179–188

Hobbs SLA, Warkentin TD, DeLong CMO (1993) Transgene copy number can be positively or negatively associated with transgene expression. Plant Mol Biol 21: 17–26

Høj PB, Hartman DJ, Morrice NA, Doan DNP, Fincher GB (1989) Purification of (1→3)-β-glucan endohydrolase isoenzyme II from germinated barley and determination of its primary structure from a cDNA clone. Plant Mol Biol 13: 31–42

Jarmolowski A, Boelens WC, Izaurralde E, Mattaj IW (1994) Nuclear export of different classes of RNA is mediated by specific factors. J Cell Biol 124: 627–635

Jorgensen R (1991) Silencing of plant genes by homologous transgenes. Agbiotech 4: 265N–273N

Kooter JM, Mol JNM (1993) Trans-inactivation of gene expression in plants. Curr Opin Biotechnol 4: 166–171

Lindbo JA, Silva-Rosales L, Proebsting WM, Dougherty WG (1993) Induction of a highly specific antiviral state in transgenic plants: implications for regulation of gene expression and virus resistance. Plant Cell 5: 1749–1759

Lotan T, Ori N, Fluhr R (1989) Pathogenesis-related proteins are developmentally regulated in tobacco flowers. Plant Cell 1: 881–887

Matzke M, Matzke AJM (1993) Genomic imprinting in plants: parental effects and trans-inactivation phenomena. Annu Rev Plant Physiol Plant Mol Biol 44: 53–76

Matzke MA, Neuhuber F, Matzke AJM (1993) A variety of epistatic interactions can occur between partially homologous transgene loci brought together by sexual crossing. Mol Gen Genet 236: 379–386

Mauch F, Mauch-Mani B, Boller T (1988) Antifungal hydrolases in pea tissue. II. Inhibition of fungal growth by combinations of chitinase and β-1,3-glucanase. Plant Physiol 88: 936–942

Meins F Jr, Neuhaus J-M, Sperisen C, Ryals J (1992) The primary structure of plant pathogenesis-related glucanohydrolases and their genes. In: Boller T, Meins F (eds) Genes involved in plant defense. Springer, Vienna, New York, pp 245–282 (Plant gene research series).

Meyer P, Heidmann I, Niedenhof I (1993) Differences in DNA-methylation are associated with a paramutation phenomenon in transgenic petunia. Plant J 4: 89–100
Mol J, Van Blokland R, Kooter J (1991) More about co-suppression. Trends Biotechnol 9: 182–183
Napoli C, Lemieux C, Jorgensen R (1990) Introduction of a chimeric chalcone synthase gene into petunia results in reversible co-suppression of homologous genes in trans. Plant Cell 2: 279–289
Neale AD, Wahleithner JA, Lund M, Bonnett HT, Kelly A, Meeks-Wagner DR, Peacock WJ, Dennis ES (1990) Chitinase, β-1,3-glucanase, osmotin, and extensin are expressed in tobacco explants during flower formation. Plant Cell 2: 673–684
Ori N, Sessa G, Lotan T, Himmelhoch S, Fluhr R (1990) A major stylar matric polypetide (sp41) is a member of the pathogenesis-related proteins superclass. EMBO J 9: 3429–3436
Sela-Buurlage MB, Ponstein AS, Bres-Vloemans SA, Melchers LS, van den Elzen PJM, Cornelissen BJC (1993) Only specific tobacco (Nicotiana tabacum) chitinases and β-1,3-glucanases exhibit antifungal activity. Plant Physiol 101: 857–863
Seymour GB, Fray RG, Hill P, Tucker GA (1993) Down-regulation of two non-homologous endogenous tomato genes with a single chimaeric sense gene construct. Plant Mol Biol 23: 1–9
Smith CJS, Watson CF, Bird CR, Ray J, Schuch W, Grierson D (1990) Expression of a truncated tomato polygalacturonase gene inhibits expression of the endogenous gene in transgenic plants. Mol Gen Genet 224: 477–481
van der Krol AR, Mur LA, Beld M, Mol JNM, Stuitje AR (1990) Flavonoid genes in petunia: addition of a limited number of gene copies may lead to a suppression of gene expression. Plant Cell 2: 291–299
Vögeli-Lange R, Fründt C, Hart CM, Beffa R, Nagy F, Meins F Jr (1994) Evidence for a role of β-1,3-glucanase in dicot seed germination. Plant J 5: 273–278
Wong Y-S, Maclachlan GA (1980) 1,3-β-D-glucanase from Pisum sativum seedlings. III. Development and distribution of endogenous substrates. Plant Physiol 65: 222–228

Gene Silencing in Transgenic Plants: A Heuristic Autoregulation Model

F. MEINS Jr. and C. KUNZ

1 Introduction

Multiple copies of transgenes introduced into plant cells can interact with one another or with homologous host genes *in trans*, resulting in the inactivation of expression of both genes (reviews: JORGENSEN 1992,1993; MATZKE and MATZKE 1993). Although the inactive state is usually stable, it is not permanent and does not result from readily discernible genetic deletions or rearrangements in DNA sequence. This phenomenon is called *silencing* (JORGENSEN 1992) or *co-suppression* (NAPOLI et al. 1990). Because co-suppression is most often applied to the interaction of transgenes with homologous host genes, the more general term silencing is used in the discussion which follows.

Silencing has many features in common with a broad range of phenomena such as paramutation (review: BRINK 1973), variation in transposon activity (FEDOROFF et al. 1989), and tumor suppression (review: MEINS 1989a) in seed plants; X-chromosome inactivation and genomic imprinting in mammals (review: SURANI 1991); and repeat-induced point mutation (RIP) (review: SELKER 1990), and methylation-induced postmeiotically (MIP) (RHOUNIM et al. 1992) in fungi. Silencing and related phenomena are of particular interest because they pose the funda-

The Friedrich Miescher Institute, P.O. Box 2543, 4002 Basel, Switzerland

mental biological problem of how cells recognize genes with the same or similar sequences *in trans*. Silencing is also of considerable practical importance. It bears directly on the interpretation of experiments with transgenes aimed at studying gene function and regulation. Moreover, applications of transgene technology for crop improvement depends on being able to control silencing to ensure reliable expression of transgenes under field conditions.

The aim of this paper is to review the salient features of transgene silencing and then to propose a heuristic model which can account for many features of silencing involving the interaction of transcribed regions of genes.

2 Properties of Transgene Silencing

Silencing of genes in transgenic plants is a general phenomenon. It has been reported for several plant species transformed with a variety of different chimeric genes (reviews: JORGENSEN 1992,1993; MATZKE and MATZKE 1993). Silencing may also be the basis for tumor suppression, i.e., the stable, reversible inactivation of genes with oncogenic functions in crown-gall disease (MEINS 1989a; AMASINO and JOHN 1989; VAN SLOGTEREN et al. 1984), and somatic reversion of the hairy-root syndrome in plants transformed with the Ri-plasmid of *Agrobacterium rhizogenes* (SINKAR et al. 1988). Silencing in the few experimental systems studied in detail has several common features:

1. Inactivation of gene expression is sequence specific, i.e., the inactivated genes have regions with the same or a very similar sequence (VAN DER KROL et al. 1990; SMITH et al. 1990; GORING et al. 1991; ASSAAD et al. 1993; MEINS and KUNZ 1994).
2. Inactivation results from a variable but substantial decrease in steady state levels of mRNA (NAPOLI et al. 1990; VAN DER KROL et al. 1990; SMITH et al. 1990; MITTELSTEN SCHEID et al. 1991; HART et al. 1992; DE CARVALHO et al. 1992; DORLHAC DE BORNE et al. 1994).
3. Silencing is often a stochastic event: only some plants in a population derived from the same transformation event show the silent phenotype and variable patterns of silencing are sometimes observed in the same plant (NAPOLI et al. 1990; VAN DER KROL et al. 1990; LINN et al. 1990; MITTELSTEN SCHEID et al. 1991; HART et al. 1992; MEYER et al. 1992; DORLHAC DE BORNE et al. 1994).
4. Once established, the silent state shows some degree of stability. It often persists for long periods during vegetative growth of the transgenic plant and, in some cases, it is transmitted through meiosis (MATZKE et al. 1989; NAPOLI et al. 1990; GORING et al. 1991; ASSAAD et al. 1993; HOBBS et al. 1993; LINN et al. 1990; MITTELSTEN SCHEID et al. 1991; MEINS and KUNZ 1994).
5. Silencing does not appear to result from permanent genetic modifications. Southern blot analyses of genomic DNA from silent plants have shown that the transgene is still present and has not undergone conspicuous rearrange-

ment or deletions (SINKAR et al. 1988; MATZKE et al. 1989; SMITH et al. 1990; LINN et al. 1990; HOBBS et al. 1990; MITTELSTEN SCHEID et al. 1991; HART et al. 1992; ASSAAD et al. 1993; DORLHAC DE BORNE et al. 1994; DEHIO and SCHELL 1994). Moreover, silent genes can undergo spontaneous as well as developmentally regulated reactivation (NAPOLI et al. 1990; MITTELSTEN SCHEID et al. 1991; HART et al. 1992; DE CARVALHO et al. 1992; MEYER et al. 1993; DORLHAC DE BORNE et al. 1994).

2.1 Sequence Specificity

Trans-inactivation of gene pairs requires some degree of sequence homology in the partners. Thus, even for host genes showing similar regulation such as the chalcone synthase gene (CHS) and the dihydroflavanol-4-reductase gene (DFR) in petunia as well as class I chitinase gene (CHN) and class I β-1,3-glucanase gene (GLU) in tobacco, introduction of the coding sequences driven by the same cauliflower mosaic virus (CaMV) 35S RNA promoter (P35S) results in the specific silencing of only the homologous host-transgene pair (VAN DER KROL et al. 1990; MEINS and KUNZ 1994). In cases where silencing involves multiple copies of the same transgene, the regions required for inactivation have not been identified. In other cases, it is clear that silencing can result from both promoter-promoter interactions and interactions between transcribed sequences. VAUCHERET (1993) has reported a striking example of promoter-dependent silencing. He transformed tobacco with T-DNA containing a neomycin phosphotransferase II gene (NPTII) regulated by the CaMV 19S RNA promoter (P19S) and terminator and a tobacco nitrite reductase gene (NIR) in reverse orientation with a double CaMV 35S RNA promoter (P35S–35S) and nopaline synthase gene (NOS) terminator. When a transformant containing this construct, which showed a Kanamycin-sensitive nitrite reductase-deficient phenotype, was crossed with tobacco plants containing a variety of different transgenes, only chimeric genes driven by P19S and P35S were inactivated, indicating that the silencing required the interaction of homologous promoter sequences. Of particular interest was the finding that only 92 bp of identical sequence was sufficient to block expression of the transgene pairs.

Related phenomena may be the complex interactions reported for doubly transformed tobacco plants by MATZKE and MATZKE and their collaborators (MATZKE et al. 1989, 1993; MATZKE and MATZKE 1991,1993). They transformed tobacco plants sequentially with two different T-DNAs. The first T-DNA (T-DNA I) contained NPTII and NOS with NOS expression signals. The second T-DNA (T-DNA II) contained the hygromycin B transferase gene (HPT) driven by CaMV 35S RNA expression signals and the octopine synthase (OCS) gene driven by the NOS promoter. Expression of NPTII and NOS was blocked in a stable but reversible fashion provided T-DNA II was present in the same genome. In this case silencing was unidirectional, i.e., genes driven by PNOS in T-DNA I but not in T-DNA II were affected.

Examples of interaction between homologous transcribed sequences include silencing of PNOS-NOS and a truncated NOS (ΔNOS) driven by the 35S promoter introduced sequentially into tobacco (GORING et al. 1991); silencing of multiple copies of P35-β-glucuronidase (GUS) in tobacco (HOBBS et al. 1990, 1993); silencing of P35S-$NIA_{tobacco}$ and the host NIA in tobacco (DORLHAC DE BORNE et al. 1994); silencing of a truncated tomato polygalacturonidase gene (ΔPG) regulated by the 35S promoter and the host PG in tomato (SMITH et al. 1990); silencing of $CHS_{petunia}$ and $DFR_{petunia}$ genes regulated by the 35S promoter and their respective host genes in petunia (NAPOLI et al. 1990; VAN DER KROL et al. 1990); and silencing of $CHN_{tobacco}$ and $GLU_{tobacco}$ genes regulated by the 35S promoter and their respective host genes in *Nicotiana sylvestris* (NEUHAUS et al. 1991; HART et al. 1992; MEINS and KUNZ 1994).

The minimum length of identical transcribed sequence required for silencing has not been established. What is clear is that silencing can result from interactions of full-length transcribed sequences with 3' truncated coding sequences (SMITH et al. 1990; GORING et al. 1991), with intron-less cDNAs, and with transcribed sequences missing either 3'-, 5'-, or both untranslated regions (VAN DER KROL et al. 1990; HART et al. 1992; DORLHAC DE BORNE et al. 1994). Taken together, these results indicate that continuous similarity along the full length of transcribed sequence is not required for silencing.

Both the transgene and homologous host genes are silenced in P35S-$CHS_{petunia}$ and P35S-$DFR_{petunia}$ transformants of petunia as well as in P35S-$CHN_{tobacco}$ and P35S-$GLU_{tobacco}$ transformants of *N. sylvestris* (NAPOLI et al. 1990; VAN DER KROL et al. 1990; MEINS and KUNZ 1994). However, sequence homology is not sufficient for silencing of a particular gene in a plant showing a silent phenotype. Thus, the homologous host gene is not silenced in P35S-$GLU_{plumbaginifolia}$ transformants of tobacco (DE CARVALHO et al. 1992) and the host phenylalanine ammonium lyase genes (PAL) but not the P35S-PAL_{bean} transgene are silenced in tobacco transformants (ELKIND et al. 1990). Similar variation has been reported for promoter-specific silencing. All genes driven by the 19S and 35S promoter and silenced by the P35S-35S-$NIR_{tobacco}$ transgene in tobacco (VAUCHERET 1993), while only the PNOS-driven transgenes in T-DNA I are silenced in double transformants of tobacco (MATZKE et al. 1989).

2.2 Gene Copy Number Effects

In several cases the incidence of silencing was found to increase with copy number (LINN et al. 1990; MITTELSTEN SCHEID et al. 1991; KUDO 1992; HOBBS et al. 1990, 1993). In other cases, no definite relationship between copy number and incidence of silencing was observed (NAPOLI et al. 1990; VAN DER KROL et al. 1990; DORLHAC DE BORNE et al. 1994; HART et al. 1992). The requirement for homozygosity at the transgene locus described for certain types of silencing may also be a gene copy effect. P35S-$CHN_{tobacco}$ and P35-β-1,3-$glucanase_{tobacco}$ transformants of *N. sylvestris* (HART et al. 1992; MEINS and KUNZ 1994), P35S-β-1,3-$glucanase_{plumbaginifolia}$

(DE CARVALHO et al. 1992) and P35S-NIA$_{tobacco}$ (DORLHAC DE BORNE et al. 1994) transformants of tobacco, and P35S-rolB transformants of *Arabidopsis* (DEHIO and SCHELL 1994) exhibit a high incidence of silencing when the plants are homozygous. Progeny obtained from crosses of these plants with wild-type plants show the high-expressing phenotype or a greatly reduced incidence of silencing. It is of interest that haploid plants regenerated from developing pollen from high-expressing P35S-β-1,3-glucanase$_{plumbaginifolia}$ transformants show the silent phenotype. This indicates that the silent state is not recessive and that this form of silencing does not depend on allelic interactions of T-DNA at the same locus (DE CARVALHO et al. 1992).

2.3 Heritability

Meiotic transmission of the silent state is variable. Silencing that depends on homozygosity at the transgene locus is not generally inherited in sexual crosses (HART et al. 1992; DE CARVALHO et al. 1992; DORLHAC DE BORNE et al. 1994). In most other cases, the silent state shows some degree of transmission in sexual crosses and appears to be inherited as a dominant trait (JORGENSEN 1993). The pattern of inheritance is often complex and consistent with the hypothesis that the silent phenotype is somewhat unstable in the hemizygous state. Thus, in crosses of a silent, homozygous line of P35S-DFR$_{maize}$ petunia transformants, which has pure white flowers, with the parent DFR-deficient line, which also has pure white flowers, ≈2% of the progeny have flowers with red variegation (MEYER et al. 1993). Reactivation of silent host and/or transgenes has been reported in backcrosses to wild type of P35S-HPT transformants of *Arabidopsis* (MITTELSTEN SCHEILD et al. 1991), P35S-CHS$_{petunia}$ transformants of petunia (NAPOLI et al. 1990), and P35-35S-NIR$_{tobacco}$ transformants of tobacco (DORLHAC DE BORNE et al. 1994). There is also evidence that in plants showing silencing, different alleles at the transgene locus can interact, resulting in a heritable change in state (MEYER et al. 1993). These phenomena show properties reminiscent of paramutation (review: BRINK 1973).

2.4 Somatic Stability

Once established, the silent state tends to persist in vegetatively growing plants. Nevertheless, individual plants can show spatial patterns of silencing. This is particularly conspicuous for CHS- and DFR-silencing, which generates deficiencies in flower pigmentation. For example, in P35S-DFR$_{maize}$ transformants of DFR-deficient petunia a significant fraction of plants hemizygous for the transgene and derived from the same primary transformant showed pale or variegated patterns of flower coloration under field conditions (MEYER et al. 1992). Some of the plants successively formed pure red, variegated, and white flowers. Similar findings have been reported for hemizygous P35S-CHS$_{petunia}$ transformants of petunia

(Napoli et al. 1990). A plant with exclusively pure-white flowers was found that gave rise to a side branch with exclusively pure-purple, i.e., wild-type, flowers. Both states of pigmentation persisted in vegetative cuttings.

Spatial patterns of silencing have also been reported for P35S-$CHN_{tobacco}$ transformants of *N. sylvestris* homozygous for the transgene. Most silent plants—≈80%—showed a uniform pattern of silencing in different leaves of the same plant. The remaining plants showed a variable pattern in which chitinase content abruptly changed between adjacent leaves (Hart et al. 1992). When plants were decapitated to promote growth of lateral shoots, the leaves that formed on these shoots continued to express the state of chitinase expression of the adjacent leaf on the parent plant (Meins and Kunz 1994).

Taken together, the results indicate that transgenic plants can exhibit stable "on" and "off" states of expression resulting in spatially uniform patterns as well as metastable or "variable" states resulting in variegation. In principle, these states could arise from epigenetic changes, i.e., mitotically transmitted, reversible alterations in phenotype, or from self-perpetuating changes at the supracellular level due to cell-cell interactions or gradients of diffusible factors (Meins 1983, 1986). The diagnostic feature of mitotically transmitted changes is that they generate clonal patterns of variegation. The few examples studied in detail favor a supracellular mechanism: the patterns of floral pigmentation generated by CHS silencing in petunia do not appear to be clonal as expected for cell-heritable variation (Napoli et al. 1990; van der Krol et al. 1990) and the patterns of P35S-CHN silencing in leaves of *N. sylvestris* do not coincide with the cell lineage for leaf formation (Hart et al. 1992). A more rigorous test for this hypothesis would be to assay numerous cloned lines derived from silent and high-expression tissues for stable expression of the two phenotypes.

2.5 Developmental and Environmental Regulation

In some cases, silencing is developmentally regulated. This is particularly striking when the silent state is not transmitted in sexual crosses and depends on homozygosity at the transgene locus. Seed of homozygous P35S-$CHN_{tobacco}$ transformants of *N. sylvestris* invariably exhibit a high-chitinase phenotype independent of the state of chitinase expression in leaves and floral organs of the parent plants (Hart et al. 1992; Meins and Kunz 1994). Young seedlings raised under conditions that promote silencing also exhibit a high-chitinase phenotype. Roughly 10 weeks post-germination, however, ≈25%–70% of the seedlings shift to the silent phenotype. These results show that chitinase-gene silencing is a stochastic process and that competence for silencing is established during a phenocritical period in seedling growth. In contrast, resetting to the high-chitinase state is not stochastic and sometimes occurs during flower development. Similar patterns of developmental regulation have been described for silencing of P35S-β-1,3-$glucanase_{plumbaginifolia}$ and P35S-$NIA_{tobacco}$ transgenes in tobacco (de Carvalho et al. 1992; Dorlhac de Borne et al. 1994), and the P35S-rolB transgene in

Arabidopsis (DEHIO and SCHELL 1994). Developmental regulation of silencing has also been reported for tomatoes hemizygous for the P35S-ΔPG transgene (SMITH et al. 1990). In ripe fruit, where the homologous host PG is normally expressed, expression of both the host gene and the transgene is markedly reduced. In unripe fruit, where the host gene is not expressed, expression of the transgene is not inhibited. This suggests that expression of the host gene is required for silencing and that developmental regulation results from a prepattern of PG expression.

The incidence of silencing often depends on environmental factors. Silencing of the P35S-$CHS_{petunia}$ transgene in petunia appears to require high light intensity (VAN DER KROL et al. 1990). The incidence of P35S-DFR_{maize} silencing is strikingly higher in field-grown plants than in greenhouse-growth plants and depends on both plant age and the age of the plants used as a source of seed (MEYER et al. 1992). Populations of P35S-$CHN_{tobacco}$ transformants of *N. sylvestris* germinated in closed culture vessels, transferred to soil, and then raised to maturity in a greenhouse show a high incidence of silencing (HART et al. 1992). When seeds are germinated in a greenhouse in open vessels containing soil, raised to maturity, and assayed, no silencing is observed. The incidence of silencing in P35S-$NIA_{tobacco}$ transformants of tobacco depends on the conditions of germination in the greenhouse and age of the plants in the field (DORLHAC DE BORNE et al. 1994).

2.6 The Site of Silencing in Gene Expression

Silencing could act at any one of the steps in gene expression, including transcription, stabilization or degradation of transcripts, translation, post-translational modification, protein turnover, and modulation of enzyme activity. The fact that inhibition of gene expression is associated with a dramatic decrease in mRNA accumulation suggests that the primary lesion is in mRNA metabolism (NAPOLI et al. 1990; VAN DER KROL et al. 1990; SMITH et al. 1990; MITTELSTEN SCHEID et al. 1991; HART et al. 1992; DE CARVALHO et al. 1992; DORLHAC DE BORNE et al. 1994). Several studies favor a transcriptional site of silencing. Methylation of cytosines in the promoter region is known to inhibit transcription of plant genes and has been proposed as a mechanism for silencing (MATZKE et al. 1989; LINN et al. 1990). This view is supported by examples in which the degree and incidence of silencing is roughly correlated with methylation of the transgene (MATZKE et al. 1989; LINN et al. 1990; MEYER et al. 1992, 1993; HOBBS et al. 1990, 1993; VAUCHERET 1993). Moreover, it was found that reactivation of silent transgenes is accompanied by decreased methylation of the gene (MEYER et al. 1993). Runoff transcription studies have shown that silencing of the P35S-DFR_{maize} transgene in petunia, which is correlated with methylation at a particular site in the 35S promoter, results from a block in transcription (MEYER et al. 1993). Blocks in transcription might also result from methylation of transcribed sequences. Recent studies with *Ascobulus* have shown that segments of methylated DNA in a transcriptionally active gene can generate truncated transcripts that appear to end at a position corresponding to the 5' end of the methylated region (BARRY et al. 1993). This provides a plausible explanation for the observation that low levels of short PG transcripts accumulate in ripe of P35S-ΔPG_{tomato} transformants of tomato showing silencing (SMITH et al. 1990).

There are examples in which no obvious correlation between methylation of the gene and silencing was found (GORING et al. 1991; MITTELSTEN SCHEID et al. 1991; HART et al. 1992). The conflicting findings are difficult to interpret because reported estimates of DNA methylation are based on comparisons of the patterns of DNA digestion obtained with restriction isoschizomers. This experimental procedure only detects methylation at a limited number of sites, which may or may not be important for transcription.

Several lines of evidence favor post-transcriptional sites of silencing. No decrease or small decreases in runoff transcription were detected in comparisons of high-expressing and silent tobacco, petunia, and *Arabidopsis* plants transformed with P35S-β-1,3-glucanase$_{plumbaginifolia}$, P35-CHS$_{petunia}$, and P35S-rolB, respectively (DE CARVALHO et al. 1992; MOL et al. 1994; DEHIO and SCHELL 1994). These results support the conclusion that initiation of transcription proceeds at the same rate in high-expressing and silent plants. It has also been proposed that silencing involves an antisense RNA mechanism in which complete or even partial antisense transcripts of the transgene are produced which in some way promote degradation of the sense transcript (SMITH et al. 1990; GRIERSON et al. 1991; MOL et al. 1991,1994). Indirect evidence for this hypothesis includes the observation that antisense—and silent sense—P35S-ΔPG_{tomato} transformants of tomato accumulate a truncated sense transcript of similar size (SMITH et al. 1990); the similarity of flower-color patterns in petunia produced by P35S-CHS$_{petunia}$ silencing and by antisense transformation with the same gene (MOL et al. 1991); and the finding of low levels of antisense CHS transcripts in sense transformants showing silencing (MOL et al. 1991, 1994).

3 An Autoregulation Model

Silencing poses several challenging problems: (1) How are genes with the same or similar sequences recognized by cells? (2) What is the nature of the *trans*-acting signal or signals? (3) How is mRNA accumulation blocked? (4) What maintains different states of gene expression in a potentially reversible fashion resulting in variable expression and variable heritability? (5) How can states with a stochastic pattern of expression be environmentally and developmentally regulated? Although, at present, there is no direct evidence that the broad range of phenomena loosely classified as silencing have the same molecular basis, the examples of developmentally regulated inactivation involving transcribed sequences are sufficiently similar to suggest, as a working hypothesis, that they have a common mechanism.

Several mechanisms for silencing have been proposed, including DNA methylation, production of antisense transcripts, ectopic gene pairing, and chromatin modification (reviews: JORGENSEN 1993; MOL et al. 1994). None is entirely satisfactory. For example, DNA methylation provides an attractive explanation for the stability and potential heritability of the inactive state and could account for effects on both the initiation and elongation steps of transcription. It does not, however, provide a solution to the more general problem of DNA sequence recognition and the nature of the *trans*-acting signal. The major strength of the antisense RNA hypothesis is that it provides a plausible explanation for sequence recognition and *trans*-inactivation, but it does not address the problem of how the production of antisense transcripts is maintained in a stable, potentially reversible fashion.

For the sake of discussion, we propose an alternative class of mechanisms based on the concept of interacting regulatory networks. These mechanisms are attractive because they can account for a variety of stable states as well as the plasticity of these states in response to developmental and environmental influences. The theory behind systems stability has been described in detail (MONOD and JACOB 1961; ROSEN 1972; MEINS 1989b; JABLONKA et al. 1992). In brief, the basic idea is that certain systems of chemical reactions and regulatory networks can exist in alternative, stable steady states. These states most commonly arise as a result of positive autoregulation, i.e., positive feedback loops in which an effector molecule either triggers its own synthesis or inhibits its own degradation. The stability and variety of phenotypes would depend on the kinetic parameters of the regulatory systems and, hence, ultimately on the genetic constitution of the cells. Thus, in principle, genetically identical cells or groups of cells can exhibit different stable states depending on their history.

There is strong evidence that positive autoregulation in which cell division-promoting factors induce their own synthesis can account for cytokinin habituation, a type of epigenetic change that occurs in cultured plant cells (MEINS and BINNS 1978; MEINS 1989b). Positive autoregulation has also been implicated in the stable, transcriptional regulation of flowering by MADS-box DNA binding proteins in *Antirrhinum* (SCHWARZ-SOMMER et al. 1992; HALFTER et al. 1994) and in the regulation of alternative states of the *Spm* transposable element in maize (FEDOROFF et al. 1989). The heuristic model we propose, which is similar to the one suggested by JABLONKA et al. (1992) for transcriptional regulation, is a modification of the model for cytokinin habituation (MEINS 1989b; MEINS and KUNZ 1994). It is limited to those forms of silencing resulting from the interaction of transcribed sequences. The model is based on three assumptions: (1) Expression of target genes with the potential of silencing is autoregulated in a sequence-specific fashion. (2) Transcription of the target gene(s) directly or indirectly results in the production of a diffusible activator of gene expression, which is degraded in a first-order process. (3) The activator interacts directly with transcribed sequences or with the transcript to increase steady state mRNA levels. The exact pathways of activator production and activator action are not specified in the model. Thus, for example, transcription of the target gene could result in signals that trigger expression of a second gene which then produces the activator.

As a concrete example, consider silencing of the P35S-CHN$_{tobacco}$ transgene and the homologous host gene in *N. sylvestris*, which is developmentally regulated and not transmitted meiotically (Hart et al. 1992). We propose that the coding region of chitinase genes binds a regulatory protein and that production of complete chitinase transcripts requires formation of a complex between this protein and the activator (Fig. 1a). Because activator production depends on transcription of both the host gene and the transgene, a positive autoregulatory loop is generated, which can exist in alternative kinetic states. Thus, the rate of activator synthesis will be a sigmoidal function of activator concentration (Fig. 1b). For purposes of illustration, the host gene is assumed to be less strongly expressed than the transgene, but with a higher affinity for the activator (X1) than the transgene (X2). Because degradation of the activator is first order, the rate of activator degradation will be directly proportional to activator concentration. Finally, because formation of the activator and chitinase mRNA is linked, the rates of chitinase mRNA and activator production from each gene will be proportional.

The net rate of activator accumulation is zero at points where the synthesis and degradation curves intersect. These points specify steady-state concentrations of activator. Where and how often the two curves intersect depend on the kinetic parameters of the system. The graph in Fig. 1b illustrates a case in which the curves intersect at five points. Three of the points specify stable states: host gene and transgene both "on," host gene "on" and transgene "off," and host gene and transgene both "off," which correspond to activator concentrations C3, C2, and C1, respectively. The three states are separated by two unstable steady states which occur at the concentrations X1 and X2. A more intuitive description of the predicted steady states is provided by the stability diagram shown at the bottom of Fig. 1b. In this representation, the three stable states are at the bottom of "valleys" and the unstable states are at the tops of "hills" (Meins 1989b).

This heuristic model can account for several important features of chitinase gene silencing, namely, sequence recognition and coding sequence specificity, which depend on the nature of the regulatory protein; *trans*-inactivation of multiple target genes, which results from the interaction of the protein with a diffusible activator; and, alternative, stable high chitinase and silent states, which result from the kinetic structure of regulatory system.

The model also provides a plausible explanation for stochastic and deterministic switching between states. Suppose that there is a momentary increase in the rate at which the activator is degraded such that the degradation and synthesis curves now intersect to give two rather than three stable states, namely, the silent host gene and transgene "off" state and the host-gene "on," transgene "off" state (Fig. 2). This could result from developmental regulation during the phenocritical period when silencing is triggered or from environmental perturbations. If the cells are initially in the high chitinase state, then when the degradation rate increases they will shift to the host gene "on," transgene "off" state. When the rate of degradation is reduced to the original level, the system

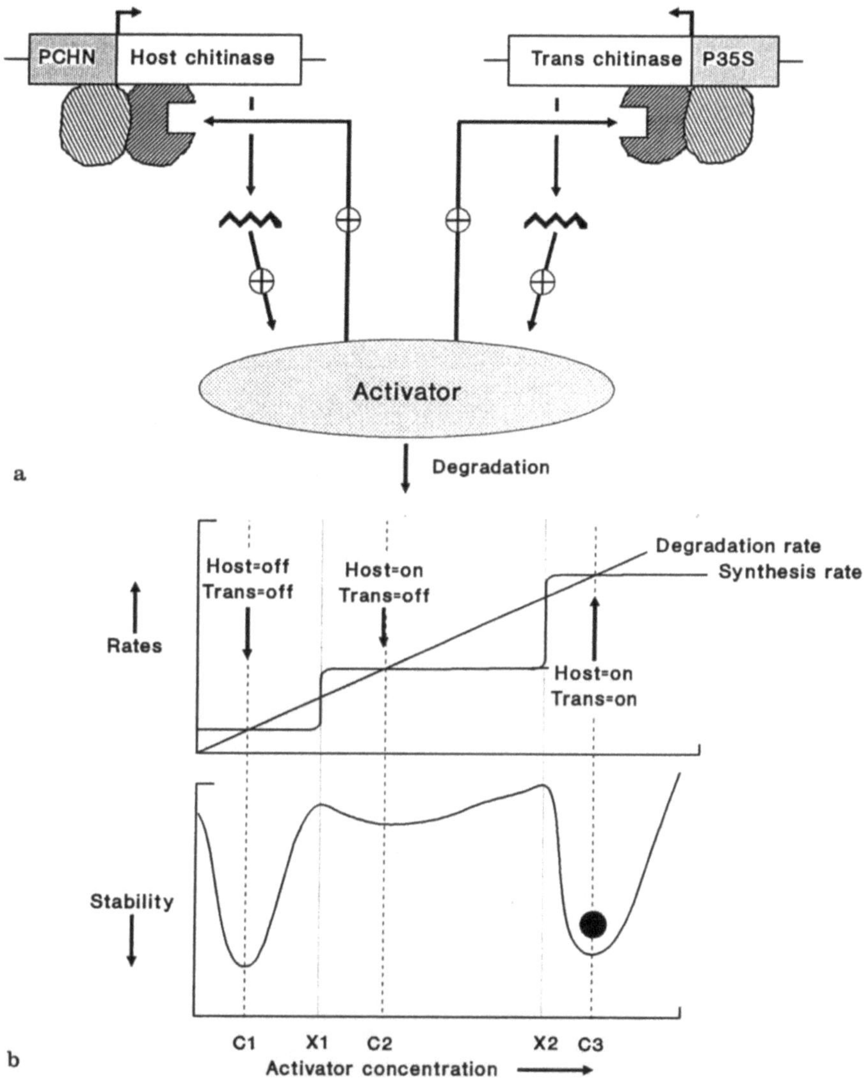

Fig. 1 a, b. A heuristic positive autoregulation model for chitinase silencing. **a** Positive autoregulation of a diffusible factor which activates chitinase gene expression by interacting with coding sequence-specific regulatory proteins to alter the rate at which complete, functional chitinase transcripts are produced. *Shaded boxes*: 5' DNA sequences regulating transcription of *N. sylvestris* (PCHN) and transgene (P35S) chitinases. **b** A kinetic representation of the model showing the stability of the different steady states. The concentrations at which stable states (*C1, C2* and *C3*) and unstable states (*X1* and *X2*) occur are indicated. (Redrawn from MEINS and KUNZ 1994)

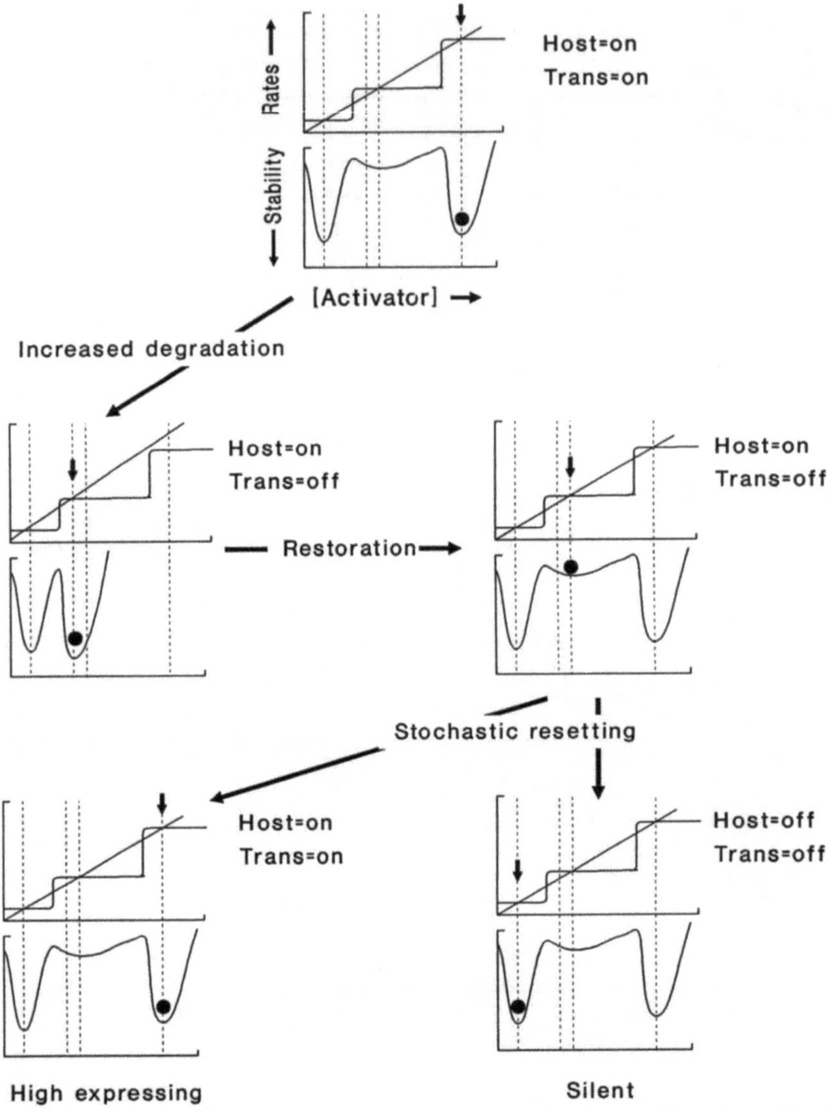

Fig. 2. A thought experiment illustrating how a momentary increase in the rate of activator degradation in response to developmental or environmental signals can result in stochastic resetting. The *vertical arrows* and *solid circles* indicate the states occupied by the system. (Redrawn from MEINS and KUNZ 1994)

will remain in this state. Inspection of the stability diagram shows that this stable state corresponds to a shallow plateau. Thus, small perturbations could, in principle, shift the system to either of the more stable high-expressing and silent states. In other words, depending upon the kinetic parameters, resetting of states could occur in a stochastic manner, even when the resetting process is initiated by a programmed, developmental event. Moreover, because resetting does not involve changes in the regulatory structure of the system, shifts between silent and high-expressing states should be fully reversible.

4 Concluding Remarks

The mechanism for silencing is not known. We suggest that autoregulation models are a useful metaphor for silencing dependent on transcribed regions of genes. These models are attractive for several reasons. First, they provide an explanation for stable, potentially reversible, developmental switching at the supragenetic level in response to diverse developmental and environmental signals. Second, they are sufficiently flexible to account for states stable at the supracellular level that are not clonally transmitted, for mitotically transmitted epigenetic changes, and for meiotically transmitted epimutations. Third, they do not specify a specific molecular mechanism. Positive autoregulation could involve transcription, as we assumed, or act at various post-transcriptional levels of regulation. Finally, they emphasize the belief that silencing is not unique to the interactions among transgenes. In principle, autoregulatory loops of the type proposed could provide a more general mechanism for stable, reversible changes important in organ determination (Meins 1988), photoperiodism (Lang 1965), and phase change (Brink 1962).

The specific model described for chitinase gene silencing is not totally unrealistic; several components of the model have already been described in plants. There is good evidence for positive autoregulation of hormone production and in the transcription of genes encoding DNA binding protein (Meins 1989c; Schwarz-Sommer et al. 1992; Halfter et al. 1994). Transcribed sequences such as introns, the protein coding region, and the 3' untranslated region contain elements known to affect steady state accumulation of plant mRNAs (Callis et al. 1987; Douglas et al. 1991; An et al. 1989; Dean et al. 1989). This model also leads to specific predictions amenable to experimental verification. For example, there should be coding sequence-specific regulation of chitinase genes, which could be tested by examining the pattern of chitinase gene expression driven by constitutive expression signals. At the systems level, specific patterns of resetting should occur when the rate of chitinase gene expression is perturbed by treatments known to regulate the host gene or by introducing into plants chimeric chitinase genes with regulated promoters.

Acknowledgements. We thank Jeff Schell and Hervé Vaucheret for providing us with unpublished information and Ortrun Mittlesten Scheid, Jean-Pierre Jost, and Jeff Dangl for stimulating discussion and criticism.

References

Amasino RM, John MC (1989) Extensive changes in DNA methylation patterns accompany activation of a silent T-DNA ipt gene in Agrobacterium tumefaciens-transformed plant cells. Mol Cell Biol 9: 4298–4303

An G, Mitra A, Choi HK, Costa MA, An K, Thornburg RW, Ryan CA (1989) Functional analysis of the 3' control region of the potato wound-inducible protease inhibitor II gene. Plant Cell 1: 115–122

Assaad FF, Tucker KL, Signer ER (1993) Epigenetic repeat-induced gene silencing (RIGS) in Arabidopsis. Plant Mol Biol 22: 1067–1085

Barry C, Faugeron G, Rossignol J-L (1993) Methylation induced premeiotically in Ascobolus: Coextension with DNA repeat lengths and effect on transcript elongation. Proc Natl Acad Sci USA 90: 4557–4561

Brink A (1962) Phase change in higher plants and somatic cell heredity. Q Rev Biol 37: 1–22

Brink RA (1973) Paramutation. Annu Rev Genet 7: 129–152

Callis J, Fromm M, Walbot V (1987) Introns increase gene expression in cultured maize cells. Genes Dev 1: 1183–1200

de Carvalho F, Gheysen G, Kushnir S, van Montagu M, Inzé D, Castresana C (1992) Suppression of β-1,3-glucanase transgene expression in homozygous plants. EMBO J 11: 2595–2602

Dean C, Favreau CM, Bond-Nutter D, Bedbrook J, Dunsmuir P (1989) Sequences downstream of translation start regulate quantitative expression of two petunia rbcS genes. Plant Cell 1: 201–208

Dehio C, Schell J (1994) Identification of plant genetic loci involved in a post-transcriptional mechanism for meiotically reversible transgene silencing. Proc Natl Acad Sci USA 91: 5538–5542

Dorlhac de Borne F, Vincentz M, Chupeau Y, Vaucheret H (1994) Co-suppression of nitrate reductase host genes and transgenes in transgenic tobacco plants. Mol Gen Genet 243: 613–621

Douglas CJ, Hauffe KD, Ites-Morales M-E, Ellard M, Paszkowski U,Hahlbrock K, Dangl JL (1991) Exonic sequences are required for elicitor and light activation of a plant defense gene, but promoter sequences are sufficent for tissue specific expression, EMBO J 10: 1767–1775

Elkind Y, Edwards R, Mavandad M, Hedrick S, Ribak O, Dixon RA, Lamb CJ (1990) Abnormal plant development and down-regulation of phenylpropanoid biosynthesis in transgenic tobacco containing a heterologous phenylalanine ammonia lyase gene. Proc Natl Acad Sci USA 87: 9057–9061

Fedoroff N, Masson P, Banks J-A (1989) Mutations, epimutations, and the developmental programming of the maize suppressor-mutator transposable element. BioEssays 10: 139–144

Goring DR, Thomson, L, Rothstein SJ (1991) Transformation of a partial nopaline synthase gene into tobacco suppresses the expression of a resident wild-type gene. Proc Natl Acad Sci USA 88: 1770–1774

Grierson D, Fray RG, Hamilton AJ, Smith CJS, Watson CF (1991) Does co-suppression of sense genes in transgenic plants involve antisense RNA? Tibtech 9: 122–123

Halfter U, Ali N, Stockhaus J, Ren L, Chua N-H (1994) Ectopic expression of a single homeotic gene, the Petunia gene green petal, is sufficent to convert sepals to petaloid organs. EMBO J 13: 1443–1449

Hart CM, Fischer B, Neuhaus J-M, Meins F Jr (1992) Regulated inactivation of homologous gene expression in transgenic Nicotiana sylvestris plants containing a defense-related tobacco chitinase gene. Mol Gen Genet 235: 179–188

Hobbs SLA, Kpodar P, DeLong CMO(1990) The effect of T-DNA copy number, position and methylation on reporter gene expression in tobacco transformants. Plant Mol Biol 15: 851–864

Hobbs SLA, Warkentin TD, DeLong CMO (1993) Transgene copy number can be positively or negatively associated with transgene expression. Plant Mol Biol 21: 17–26

Jablonka E, Lachmann M, Lamb MJ (1992) Evidence, mechanisms and models for the inheritance of acquired characters. J Theor Biol 158: 245–268

Jorgensen R (1992) Silencing of plant genes by homologous transgenes. AgBiotech News Info 4: 265N–273N

Jorgensen R (1993) The germinal inheritance of epigenetic information in plants. Phil Trans R Soc Lond [Biol] 339: 173–181

Lang A (1965) Progressiveness and contagiousness in plant differentiation and development. Encycl Plant Physiol 15(1): 409–423

Linn F, Heidmann I, Saedler H, Meyer P (1990) Epigenetic changes in the expression of the maize A1 gene in Petunia hybrida: role of numbers of integrated gene copies and state of methylation. Mol Gen Genet 222: 329–336

Matzke MA, Matzke AJM (1991) Differential inactivation and methylation of a transgene in plants by two suppressor loci containing homologous sequences. Plant Mol Biol 16: 821–830

Matzke MA Matzke AJM (1993) Genomic imprinting in plants: parental effects and trans-inactivation phenomena. Annu Rev Plant Physiol Plant Mol Biol 44: 53–76
Matzke MA, Primig M, Trnovsky J, Matzke AJM (1989) Reversible methylation and inactivation of marker genes in sequentially transformed tobacco plants. EMBO J 8: 643–649
Matzke MA, Neuhuber F, Matzke AJM (1993) A variety of epistatic interactions can occur between partially homologous transgene loci brought together by sexual crossing. Mol Gen Genet 236: 379–386
Meins F Jr (1983) Heritable variation in plant cell culture. Annu Rev Plant Physiol 34: 327–346
Meins F Jr (1986) Determination and morphogenetic competence in plant tissue culture. Bot Monogr 23: 7–25
Meins F Jr (1988) Cell commitment and determination in plants. Genet Eng 10: 141–153
Meins F Jr (1989a) Tumor reversal and tumor suppression in plants. In: Herberman RB (ed) Influence of the host on tumor development, Kluwer Academic, Dordrecht, pp 37–41
Meins F Jr (1989b) A biochemical switch model for cell-heritable variation in cytokinin requirement. In: Goldberg R (ed) Molecular basis of plant development, Liss, New York, pp 13–24
Meins F Jr (1989c) Habituation: heritable variation in the requirement of cultured plant cells for hormones. Annu Rev Genet 23: 395–408
Meins F Jr, Binns AN (1978) Epigenetic clonal variation in the requirement of plant cells for cytokinins. In: Subtelny S, Sussex IM (eds) The clonal basis for development. Academic, New York, pp 185–201
Meins F Jr, Kunz C (1994) Silencing of chitinase expression in transgenic plants: an autoregulatory model. In: Paszkowski J (ed) Homologous recombination in plants. Kluwer Academic, Dordrecht, pp 335–348
Meyer P, Linn F, Heidmann I, Meyer z.A. H, Niedenhof I, Saedler H (1992) Endogenous and environmental factors influence 35S promoter methylation of a maize A1 gene construct in transgenic petunia and its colour phenotype. Mol Gen Genet 231: 345–352
Meyer P, Heidmann I, Neidenhof I (1993) Differences in DNA-methylation are associated with a paramutation phenomenon in transgenic petunia. Plant J 4: 89–100
Mittelsten Scheid O, Paszkowski J, Potrykus I (1991) Reversible inactivation of a transgene in Arabidopsis thaliana. Mol Gen Genet 228: 104–112
Mol J, van Blokland R, Kooter J(1991) More about co-suppression. Tibtech 9: 182–183
Mol JNM, van Blokland R, de Lange P, Stam M, Kooter JM (1994) Post-transcriptional inhibition of gene expression: sense and antisense genes. In: Paszkowski J (ed) Gene inactivation and homologous recombination in plants. Kluwer Academic, Dordrecht, pp 309–334
Monod J, Jacob F (1961) General conclusions: telenomic mechanisms in cellular metabolism, growth and differentiation. Cold Spring Harb Symp Quant Biol 26: 389–401
Napoli C, Lemieux C, Jorgensen R (1990) Induction of a chimeric chalcone synthase gene into petunia results in reversible co-suppression of homologous genes in trans. Plant Cell 2: 279–289
Neuhaus J-M, Ahl-Goy P, Hinz U, Flores S, Meins F Jr (1991) High-level expression of a tobacco chitinase gene in Nicotiana sylvestris. Susceptibility of transgenic plants to Cercospora nicotianae infection. Plant Mol Biol 16: 141–151
Rhounim L, Rossignol J-L, Faugeron G (1992) Epimutation of repeated genes in Ascobulus immersus. EMBO J 11: 4451–4457
Rosen R (1972) Mechanics of epigenetic control. In: Rosen R (ed) Foundations of mathematical biology, 2nd edn. Academic, New York, pp 79–140
Schwarz-Sommer Z, Hue I, Huijser P, Flor PJ, Hansen R, Tetens F, Loenning WE, Saedler H, Sommer H (1992) Characterization of the Antirrhinum floral homeotic MADS-box gene deficiens. Evidence for DNA binding and autoregulation of its persistent expression throughout flower development. EMBO J 11: 251–264
Selker EU (1990) Premeiotic instability of repeated sequences in neurospora crassa. Annu Rev Genet 24: 579–613
Sinkar VP, White FF, Furner IJ, Abrahamsen M, Pythoud F, Gordon MP (1988) Reversion of aberrant plants transformed with Agrobacterium rhizogenes is associated with the transcriptional inactivation of the T_L-DNA genes. Plant Physiol 86: 584–590
Smith CJS, Watson CF, Bird CR, Ray J, Schuch W, Grierson D (1990) Expression of a truncated tomato polygalacturonase gene inhibits expression of the endogenous gene in transgenic plants. Mol Gen Genet 224: 477–481
Surani MA (1991) Genomic imprinting: developmental significance and molecular mechanism. Curr Open Genet Dev 1: 141–246

van der Krol AR, Mur LA, Beld M, Mol JNM, Stuitje AR (1990) Flavonoid genes in petunia: addition of a limited number of gene copies may lead to a supression of gene expression. Plant Cell 2: 291–299
van Slogteren GMS, Hooykaas PJJ, Schilperoort RA (1984) Silent T-DNA genes in plant lines transformed by Agrobacterium tumefaciens are activated by grafting and by 5-azacytidine treatment. Plant Mol Biol 3: 333–336
Vaucheret H (1993) Identification of a general silencer for 19S and 35S promoters in a transgenic tobacco plant : 90 bp of homology in the promoter sequence are sufficient for trans-inactivation. C R Acad Sci (III) 316: 1471–1483

Paramutation in Maize and Related Allelic Interactions

G.I. Patterson[1,2] and V.L. Chandler[1]

1 Introduction

Paramutation has been defined as an interaction between alleles that leads to a mitotically and meiotically heritable change in one of the alleles at high frequency (Brink 1973). In the cases in which the gene function is known, it is clear that paramutation leads to reduction but not loss of phenotypic expression of the affected allele. We will discuss the genetic properties of paramutation of the maize genes *b, r,* and *pl* and review recent characterization of the effect of paramutation on gene expression and structure. As we review paramutation, we will compare apparently related phenomena, including: (1) recently described interactions between transgenes and endogenous genes in plants that lead to gene silencing, (2) interactions between transposable elements of plants that lead to heritable changes in gene expression, and (3) transvection in *Drosophila*

[1] Institute of Molecular Biology, University of Oregon, Eugene, OR 97403, USA
[2] Current address: Department of Molecular Biology, Massachusetts General Hospital, Boston, MA 02114, USA

and heritable epigenetic effects on gene expression in animals and fungi. For a thorough review of earlier literature on paramutation of *b, r,* and the handful of other examples of paramutation in plants, we refer the reader to previous reviews (Brink 1964, 1973; Brink et al. 1968; Coe 1966).

Paramutation has been characterized genetically at the *b, r,* and *pl* loci in maize (Coe 1966; Brink et al. 1968; Patterson et al. 1993a). *b, r, c1,* and *pl* regulate the synthesis of red and purple anthocyanin pigments by regulating transcription of the anthocyanin biosynthetic genes (Goff et al. 1990; Roth et al. 1991; Bodeau and Walbot 1992). Genetic studies showed that certain alleles of the *b* and *r* genes are functionally equivalent (Styles et al. 1973). Sequencing of cDNA clones showed that *b* and *r* are homologous and share sequence identify with transcriptional activators of the basic helix-loop-helix class (Ludwig et al. 1989; Radicella et al. 1991). Similarly, *c1* and *pl* are functionally equivalent (Cocciolone and Cone 1993), are homologous, and share sequence identity with transcriptional activators of the myb class (Paz-Ares et al. 1987; Cone et al. 1993). One member of each of the two classes of regulatory genes function together to activate transcription of anthocyanin biosynthetic genes (Goff et al. 1992).

2 Genetics of Paramutation

2.1 Paramutation Is a Directed, Meiotically Heritable Change in an Allele

Paramutation of *b* was first described by Coe (1966), and is illustrated by showing the results of a set of crosses (Fig. 1a). Homozygous *B-I/B-I* plants are intensely purple because anthocyanin pigment accumulates at high levels in a number of tissues. Homozygous *B'/B'* plants accumulate less anthocyanin than *B-I/B-I* plants, but in the same tissues. The *B'/B-I* phenotype is similar to that of *B'/B'*. This light phenotype is unusual in that *B-I* is typically dominant to *b* alleles that produce lower levels of pigment (see control cross, Fig. 1a). Remarkably, when a *B'/B-I* plant is crossed to a *b-K55/b-K55* tester, all progeny are *B'/b-K55*, rather than one half *B-I/b-K55* and one half *B'/b-K55*, as would be expected (Fig. 1a). When *B-I/B-I* is crossed to *b-K55/b-K55* as a control the plants are dark. Thus, *B'* changes *B-I* into *B'* in the heterozygote. This alteration, paramutation, is stable through meiosis. In this review, we will follow the convention of referring to *B'* and *B-I* as different alleles of b; however, it is not clear whether *B'* and *B-I* are different alleles or different epigenetic states of the same allele.

A number of genetic parameters of the *b* paramutation phenomenon have been elucidated. The original *B'* allele arose as a spontaneous derivative of *B-I*. The failure to segregate in the F2 (Fig. 1a) is not due to segregation distortion, as markers linked to *B-I* segregate normally (Coe 1966). There is no evidence for parental imprinting; the phenotypes seen do not depend on whether *b* alleles

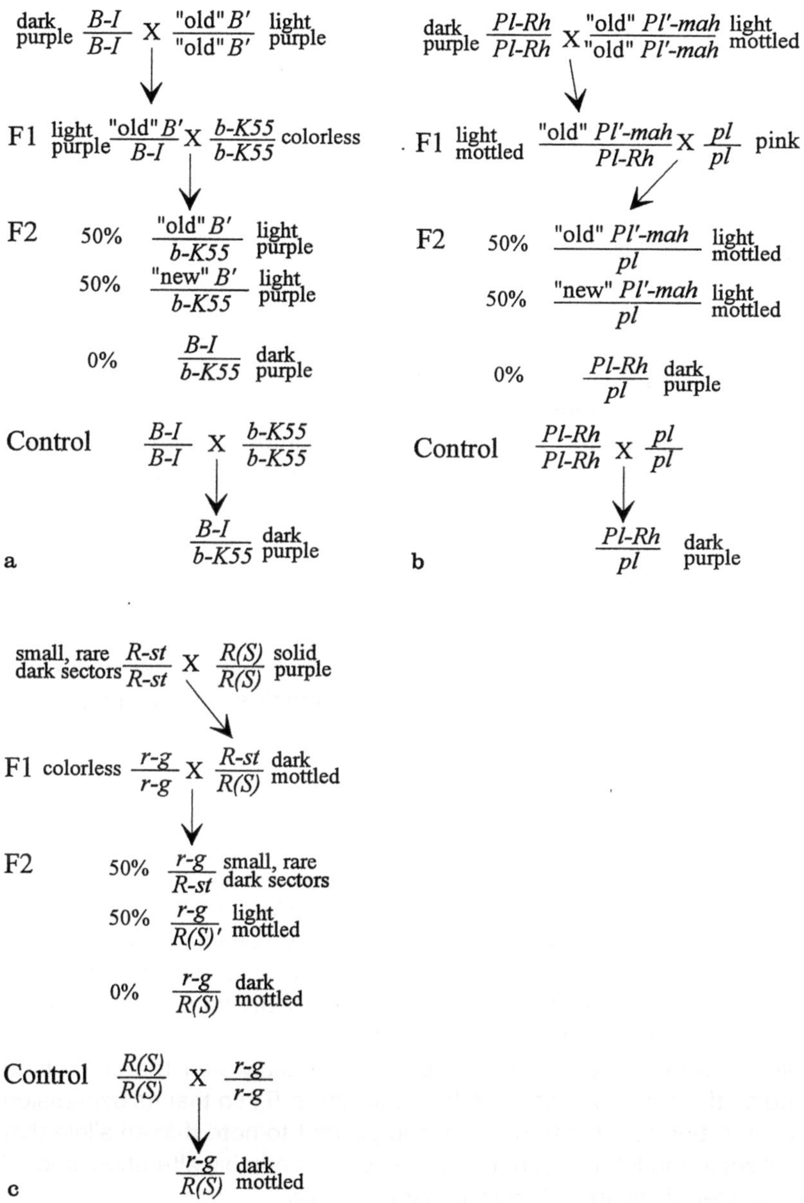

Fig. 1a–c. Summary of paramutation. See Sect. 2.1 for details. **a** *b* gene paramutation. The *b-K55* allele is nonfunctional: it produces no pigmentation, is not affected by *B'*, and does not affect *B-I*. The "new" *B'* shown in the F2 is the allele that was previously *B-I*. **b** *pl* gene paramutation. The "new" *Pl'-mah* allele is the allele that was previously *Pl-Rh*. **c** *r* gene complex paramutation

are transmitted by the male or female (COE 1966). Six *b* alleles (*B-bar*, *B-W22*, *B-Bolivia*, *B-Peru*, *b-K55*, and a non-functional *b* allele of uncertain derivation), with different tissue-specific patterns of anthocyanin (COE 1979), have been tested for paramutation. Only *B′* and the closely related *B′-v* and *b′-v* alleles paramutate *B-I*, and only *B-I* and the closely related *B-v* and *b-v* alleles are altered by *B′* (COE 1966; COE et al. 1988; G. Patterson and V. Chandler, unpublished observations). *B-I* spontaneously changes to *B′* at a high frequency. In the authors' stocks with a W23 genetic background, a few percent of the progeny of a *B-I/B-I* plant receive the *B′* allele. In contrast, *B′* is very stable. Efforts to select for dark plants in lineages with *B′* did not result in the identification of reversion to *B-I* (COE 1966). No revertants of *B′* to *B-I* were seen in ~10^5 progeny of a *B′/B′* by *B-I/B-I* cross (K. Kubo, G. Patterson and V. Chandler, unpublished observations).

Paramutation can be divided into three separate but perhaps related events, which are described below for *b* paramutation. First, in the initial *B′/B-I* heterozygote, very little pigment accumulates. Thus, either *B′* has already changed *B-I* into *B′* by the late developmental stage at which expression normally occurs, or *B′* suppresses *B-I* expression. Second, *B′* heritably alters the phenotype of *B-I*; the "new" *B′* (in the F2), which was *B-I* previously, produces weak pigment, as does the original *B′* ("old" *B′*). Third, the "new" *B′* gains the capacity to invariably change a naive *B-I* allele (i.e., a *B-I* allele not previously exposed to *B′*) into *B′*. These changes may be mechanistically related, as all three occur under the influence of *B′* in the *B′/B-I* heterozygote. The following three assays are used to measure paramutation:

1. *F1 phenotype:* Does the presence of an inducing allele decrease phenotypic expression of the sensitive allele in the F1? The F1 assay is useful because it is the quickest to perform, and, in all gene systems in which paramutation is seen, alleles that cause heritable alteration in sensitive alleles show at least some effect in the initial heterozygote (BRINK 1973; COE 1966; MCWHIRTER and BRINK 1963; BRINK et al. 1970; G. Patterson and V. Chandler, unpublished observations). *B′* has a very strong effect on *B-I* in an initial heterozygote (COE 1966), and the faint pigmentation in *B′/B-I* is due to low transcription of both alleles (PATTERSON et al. 1993b; see Sect. 3.1).
2. *Heritable alteration:* Does the inducing allele heritably alter the phenotype produced by the sensitive allele? *B′* heritably alters *B-I* so that its expression is almost indistinguishable from *B′*. It is important to note that an allele that has an effect in the F1 assay, but fails to cause a heritable alteration, should not be considered capable of causing paramutation.
3. *Secondary paramutation:* Does the inducing allele confer upon the sensitive allele the ability to heritably alter another sensitive allele? As described above, a "new" *B′* invariably paramutates *B-I*, and pigment phenotype of *B-I* after exposure to "new" *B′* is similar to that of *B-I* after exposure to "old" *B′*.

pl paramutation occurs when the *Pl-Rh* allele is heterozygous with the *Pl′-mah* allele. As is the case with *b* paramutation, *Pl′-mah* has a strong effect as measured by all three assays for paramutation (PATTERSON et al. 1993a,

G. Patterson and V. Chandler, unpublished observations.) As shown in Fig. 1b, the phenotype of a *Pl'-mah/Pl-Rh* heterozygote is light, like that of a *Pl'-mah/Pl'-mah* plant. When the heterozygote is crossed to a recessive *pl* allele, the phenotype conferred by a "new" *Pl'-mah* is light, and not obviously different from that of "old" *Pl'-mah.* The phenotype of *Pl'-mah* is different from *Pl-Rh* in all tissues in which color is induced by *pl,* but the "light mottled" phenotype mentioned in Fig. 1b is that of anthers and seedling sheath, two tissues in which the phenotype is the most obvious. Finally, *pl* shows strong secondary paramutation; the "new" *Pl'-mah* is 100% efficient at paramutating a naive *Pl-Rh* allele. As with *b* paramutation, the inducing allele (*Pl'-mah*) can arise as a spontaneous derivative of the sensitive allele (*Pl-Rh*). In some of the authors' stocks, several percent of the progeny of *Pl-Rh/Pl-Rh* receive *Pl'-mah.*

In *r* paramutation, as in *b* paramutation, an allelic interaction induces a heritable change in the pigmentation potential of a sensitive allele (reviewed in BRINK 1964). *r* is a complex locus that contains multiple homologous *r* genes[1]. In *r* paramutation, several inducing *r* gene complexes are known, including *R-st* (Fig. 1c). *R(S)* ordinarily produces dark, mottled pigmentation in the aleurone of an *r-g/R(S)* individual (see control cross in Fig. 1c). After being heterozygous with *R-st, R(S)* is altered to *R(S)'*, and produces very light, mottled pigmentation in the aleurone of an *r-g/R(S)'* individual (Fig. 1c). The quantitative difference in pigment produced by *R(S)* and *R(S)'* is about sevenfold (VAN DER WALT and BRINK 1969).

Comparison of *b, r* and *pl* paramutation illustrates a few key differences. First, *r* is unique in that the strength of the response of *R(S)* to paramutation depends on the assay used. In the F1 assay, *R(S)* is very weakly responsive to *R-st*; in fact, response is only seen when unusually sensitive alleles are used (MCWHIRTER and BRINK 1963; BRINK et al. 1970). In the secondary paramutation assay, *R(S)'* does alter *R(S),* but its effect is barely detectable (BROWN and BRINK 1960). *B-I* and *Pl-Rh* respond strongly, and to the same extent, in all three assays. A second difference is that *r* and *pl* show a high level of quantitative variation not seen at *b. R(S)'* and *Pl'-mah* can revert and produce darker pigmentation, but typically do not revert completely to their original state (BRINK 1964; STYLES and BRINK 1969; G. Patterson,

---

[1] Notes on maize nomenclature : (1) The locus traditionally referred to as *R-r* is a gene complex with two genes (STADLER and NUFFER 1953; DOONER and KERMICLE 1971; ROBBINS et al. 1991), and following the nomenclature of DOONER et al. (1991), we refer to these genes as *R(S)* and *R(P).* Fine structure mapping of the *r* locus has made it clear that it is the *R(S)* component that is responding to paramutation, and that *r* complexes that contain only *R(S)* and not *R(P)* respond the same way to *R-st* as complexes that contain both components (BROWN 1966). A variety of sensitive *r* gene complexes were used in the studies reviewed here, and all had the *R(S)* component, so we will use the *R(S)* designation as a generic term for the sensitive alleles used, and *R(S)'* as a generic term for the allele that results after paramutation. (2) Maize is diploid, but the aleurone layer of the seed, where *R(S)'* and *R(S)* are expressed, is triploid, having received one dose of *r* from the male parent and two doses from the female parent. For simplicity, we will refer to the diploid *r* genotype when describing genotypes. Almost invariably, *R(S)* and *R(S)'* are scored when transmitted through the male, since *r* is an imprinted gene (reviewed in KERMICLE and ALLEMAN 1990). In every case in this review, when we refer to the aleurone of an *r-g/R(S)'* plant, or the phenotype produced by *R(S)'*, we are referring to an aleurone with the triploid genotype *r-g/r-g/R(S)'*.

Table 1. Comparison of *b, r,* and *pl* paramutation

Features of paramutation[a]	*b*	*r*	*pl*
Response in three assays			
F1 pigmentation	Yes	–[b]	Yes
Heritable alteration	Yes	Yes	Yes
Secondary paramutation	Yes	–[b]	Yes
Spontaneous paramutant derivatives	Yes	–[c]	Yes
Reversion of paramutated allele	No	Yes[d]	Yes[d]
Phenotype of paramutated allele	Uniform[e]	Variable	Variable
Strength of inducing alleles	Uniform	Variable	Variable
Transcriptional repression	Yes	Not determined	Yes
Simple locus	Yes	No	Yes
Correlation with hypermethylation	None dectected	Yes	Not determined

[a] These featurers of paramutation are described in the text, and references are given there
[b] *R(S)* responds to *R-st* very weakly in these two assays
[c] *R(S)* spontaneously undergoes only very small changes, as described in the text
[d] *Pl'-mah* and *R(S)'* may not revert completely
[e] Different *B'* isolates show very slight variation (not more than twofold), wherease *R(S)'* and *Pl'-mah* vary over a much larger range

J. Hollick and V. Chandler, unpublished observations). *B'* is stable, but *R(S)'* can exist in a virtual continuum of heritable and unstable genetic states, producing dark to almost undetectable pigment (Brink et al. 1968; Styles and Brink 1969). Preliminary experiments indicate that *Pl'-mah* shows variability similar to *R(S)'* (G. Patterson, J. Hollick and V. Chandler, unpublished observations). A third difference is that both *B-I* and *Pl-Rh* spontaneously change to *B'* or *Pl'-mah* in homozygous plants (Coe 1966; Patterson et al. 1993a). *R(S)* can spontaneously become more *R(S)'* -like, but the resulting changes in color and ability to induce paramutation are very small (Brink et al. 1968). These features, and others described in Sects. 3 and 4, are summarized in Table 1. It may be that the fundamental change in gene structure that underlies these examples of paramutation is the same, and that the differences in phenomenology reflect subtle differences in the alleles. Alternatively, it may be that these differences in phenomenology indicate fundamentally dissimilar mechanisms. Until the mechanism of paramutation is elucidated, it is important when thinking about paramutation, to recognize the significant differences between the phenomena at *b, r,* and *pl*.

2.2 Paramutation Is an Allelic Interaction

Coe (1966) demonstrated that the ability of *B'* to paramutate *B-I* is genetically linked to *B'*. He made a *B'/b-K55* heterozygote; *b-K55* is an allele of *b* that is nonfunctional for activating the anthocyanin pathway and incapable of paramutating *B-I* or responding to *B'*. Progeny of a *B'/b-K55* by *B-I/B-I* cross were examined for the ability to paramutate as measured by the F1 pigmentation and heritable alteration assays (see Sect. 1.3). All of the progeny met the expectation for linkage; all *b-K55/B-I* individuals were dark, and all *B'/B-I* individuals were light (~50 of each genotype tested). *B-I* was heritably altered only in the *B'/B-I* individuals.

A larger test for linkage was done with *B'/B-Peru* heterozygotes (PATTERSON and CHANDLER, manuscript in preparation). *B-Peru* colors the seed and some tissues in the mature plant (STYLES et al. 1973; PATTERSON et al. 1991), so can be distinguished from *B'* and *B-I*. *B-Peru* does not paramutate *B-I* or respond to *B'* (COE et al. 1988; G. Patterson and V. Chandler, unpublished observations). *B'/B-Peru* plants were crossed to *B-I/B-I* plants and colorless seeds (*B'/B-I*) were separated from purple seeds (*B-Peru/B-I*) and planted. All of the 3058 resulting *B'/B-I* plants were light, demonstrating that the ability to paramutate always segregated with the regulatory sequences that provide the *B'* tissue-specific pattern of expression. While the possibility of a very tightly linked gene causing paramutation cannot be eliminated by these data, this result supports the hypothesis that the sequences involved in paramutation are a part of the *B'* allele.

Similar, small-scale experiments showed that the ability to paramutate *R(S)* is linked to *R-st* (BRINK 1958), and that the ability to paramutate *Pl-Rh* is linked to *Pl'-mah* (PATTERSON et al. 1993a). Intragenic recombination experiments (Sect. 3.2) show that the ability to paramutate *R(S)* is intimately associated with the *r* gene complex.

3 Paramutation and Gene Expression

3.1 Transcriptional Changes

The mechanism of *b* paramutation was investigated by testing for changes in *b* transcription and function (PATTERSON et al. 1993b). Transcription assays using isolated nuclei indicated that *b* transcription was lower in *B'/B'* and *B'/B-I* husks than in *B-I/B-I* husks by about tenfold, but no change was observed in the tissue-specific pattern of *b* RNA accumulation, nor were the size and 5' end of *b* RNA altered. Similarly, RNA accumulation and transcription are lower in *Pl'-mah/Pl-Rh* tissues than in *Pl-Rh/Pl-Rh* (G. Patterson and V. Chandler, unpublished observations). It should be noted that the *B'* and *Pl'-mah* alleles are not silent; transcription is still readily detected. The activity of the *B'* and *B-I* proteins was not different, as measured by their ability to activate the transcription of the target genes *bz1*, *a1* and *c2*. COE (1966) showed that the transcriptional activator function of *b* is not required for paramutation. He used a *b-v* derivative of *B-I*, which has a transposable element-induced mutation in the *b* coding region that renders it defective for activation of the anthocyanin pathway. This allele is still transcribed (COE 1966; G. Patterson, T. Shroyer and V. Chandler, unpublished observations). The *b-v* allele can participate in paramutation. When *b-v/B'* is crossed to *B-I/B-I*, two types of progeny are seen: *b'-v/B-I* and *B'/B-I*. *B-I* is paramutated in *b'-v/B-I*, that is, it is changed to *B′* (COE 1966). Thus, autogenous regulation of *B′/B-I* can only be involved in paramutation if the autogenous regulation does not require the same *b* function by which it regulates anthocyanin biosynthetic genes.

The question of whether transcriptional or post-transcriptional changes are occurring is significant for the evaluation of models. Paramutation shares some features with recently described *trans*-inactivation phenomena in plants, in which transgenes and endogenous genes that share homologous sequences interact (reviewed in MATZKE and MATZKE 1993). Models that involve physical contact between alleles or transgenes resulting in transcriptional silencing have been proposed. This type of model is adequate for *b* and *pl* paramutation, and for two cases of *trans*-inactivation in which transcription is affected (MEYER et al. 1993; BRUSSLAN et al. 1993). In three cases of *trans*-inactivation that are not heritable (DE CARVALHO et al. 1992; VAN DER KROL et al. 1990; ELKIND et al. 1990), inactivation is not accompanied by a decrease in transcription rates in isolated nuclei (DE CARVALHO et al. 1992; MOL et al. 1991; C. Lamb, personal communication). This result suggests that in these cases, *trans*-inactivation is likely to be the result of post-transcriptional events. Models involving the inadvertent production of antisense RNA have been proposed for cases in which regulation is post-transcriptional (GRIERSON et al. 1991). In one case, the transgene produced antisense RNA that was homologous to the inactivated gene (MOL et al. 1991). Antisense RNA has been shown to decrease RNA accumulation in transgenic plants (ROTHSTEIN et al. 1987).

3.2 Sequences Involved in Paramutation

The sequences necessary for the ability to respond to and cause paramutation are likely to include transcriptional regulatory sequences, as the phenomenon is associated with reduced transcription and maps to the loci. However, other sequences may be involved as well. The strategy used for mapping sequences necessary for *b* paramutation is modelled after that described by KERMICLE (1974) and is shown in Fig. 2. *B′* and *B-Peru* differ in ability to paramutate and in tissue-specific pattern of expression (Fig. 2). These alleles and mutant derivatives were used to isolate hybrid alleles by intragenic recombination. These alleles were then tested for the tissue-specific pattern of anthocyanins induced, and for the ability to paramutate (Patterson and Chandler, manuscript in preparation).

The location of sequences controlling tissue-specific gene expression in *B′* and *B-Peru* maps to the 5' region of the gene. A 2.1-kb DNA fragment from *B-Peru* containing the promoter and untranslated leader is sufficient to confer expression in the seed in transient transformation assays, whereas a 1.5-kb DNA segment from the same region of *B-I* does not promote seed expression (RADICELLA et al. 1992). Since the sequences of *B′* and *B-I* are the same in this 1.5-kb segment, this experiment also demonstrates that the 5' region of *B′* does not promote seed expression. Consistent with these transformation results, *BP//B′* recombinants exhibited a *B-Peru* pattern of tissue-specific expression in both seed and plant,

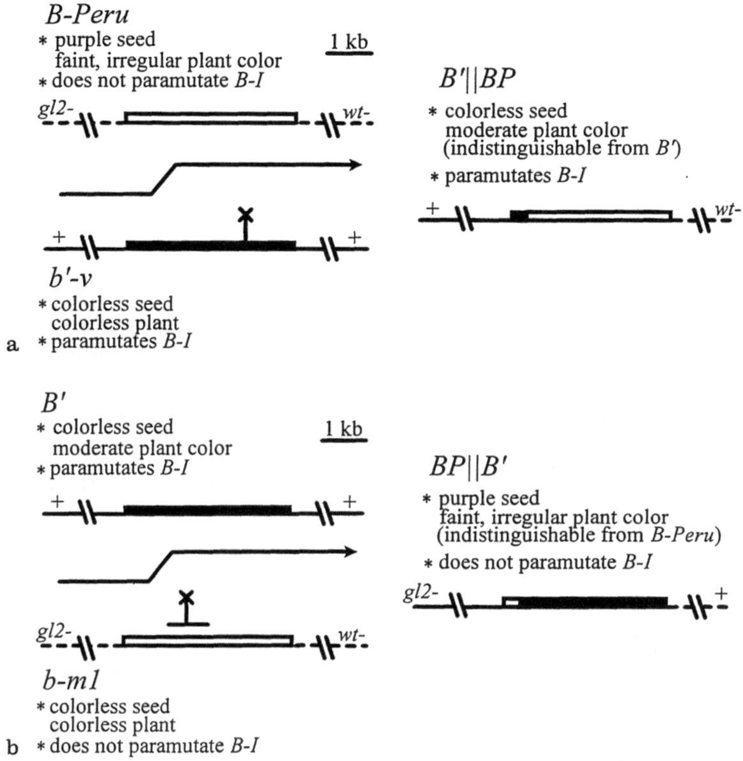

Fig. 2 a, b. Isolation and phenotypes of *b* intragenic recombinants. **a** Intragenic recombinant alleles that are *B'* upstream and *B-Peru* downstream were isolated by making a heterozygote between *B-Peru* and *b'-v* (a green mutant derivative of *B'* that paramutates *B-I* with the same efficiency as *B'*). The transcribed regions of the genes are shown as boxes. The location of the lesion in *b'-v* is shown as an *X*. Progeny of this heterozygote were screened for the presence of rare recombinant alleles that confer a *B'* color phenotype, and these alleles were tested for the ability to paramutate. *gl2* and *wt* are flanking morphological markers used to follow recombination. **b** Intragenic recombinants that are *B-Peru* upstream and *B'* downstream were isolated by making heterozygotes between *B'* and *b-m1* (a mutant derivative of *B-Peru*; CLARK et al. 1990). The lesion in *b-m1* is shown with an *X*; the insertion is mapped to a restriction fragment that is indicated by a *line above the gene.* Progeny were screened for rare recombinant alleles that confer a *B-Peru* color phenotype, and these alleles were tested for ability to paramutate

and *B'//BP* recombinants exhibited a *B'* pattern of expression (Fig. 2) These results indicate that the 5' regions of *B'* and *B-Peru* control expression in plant tissues.

The ability of *B'* to paramutate *B-I* also maps to the 5' region of the *B'* allele. *B'//BP* recombinants, which have *B'* 5' sequences, efficiently paramutate *B-I*, both in F1 pigmentation assays and in heritable alteration assays (see Sect. 1.3 for a description of the assays). In contrast *BP//B'* intragenic recombinants do not paramutate *B-I*. As described in Sect. 2.2, 3058 plants were scored in an attempt to identify a recombination event in which the ability to paramutate was separated from the sequences controlling *B'* tissue-specific expression by recombination.

No separation was found; therefore, this ability is not more than 0.1 cM from *B'*. The unique ability of *B'* to paramutate *B-I* resides within the 5' region of the allele, or in closely linked sequences upstream of the transcription start site. It is important to note that the *B-Peru* and *B'* alleles differ in their 5' flanking regions and show no sequence identity for at least 2.5 kb pairs (Radicella et al. 1992). Probes from *B'* do recognize sequences further upstream in *B-Peru*, but details of the structure are not known. These mapping data, along with the fact that paramutation affects *b* transcription, suggest the interesting possibility that sequences that are responsible for paramutation may overlap or be identical to sequences that are responsible for the transcriptional regulation. However, it is equally possible that additional sequences responsible for paramutation are genetically linked but not coincident with transcriptional regulatory sequences. The identification of sequences important for transcriptional regulation and paramutation by DNA-mediated transformation and mutagenesis is essential for resolving this question.

The location of the sequences that control paramutation is clearly very different for *b* and r. The *R-st* gene complex has been fine-structure mapped and consists of a chromosomal region with multiple linked *r*-homologous genes (W. Eggleston and J. Kermicle, personal communication). The ability of *R-st* to paramutate the sensitive *R(S)* is distributed across this chromosomal region (J. Kermicle and W. Eggleston, personal communication; Ashman 1970; Kermicle 1970a). Unequal crossing over has produced derivatives in which *r* genes are deleted, along with all intervening DNA. The ability of these *R-st* derivatives to paramutate *R(S)* is decreased in steps with each deletion. When only one gene remains, the remainder of the complex is not able to alter *R(S)* in a detectable way. Since the deletions remove one or more *r* coding regions, their associated regulatory regions, and undetermined amounts of flanking DNA, the ability to paramutate might reside in any of these sequences.

The *b* and *pl* genes do not have the composite nature of *r*. *b* and *pl* consist of only one gene (Patterson et al. 1991; Cocciolone and Cone 1993; Cone et al. 1993). The involvement of repetitive sequences in paramutation is difficult to test because maize genes are surrounded by repetitive sequences. However, the recombination analysis presented above does not indicate that *b* paramutation can be lost in a stepwise fashion; *BP//B'* is completely unable to paramutate *B-I*, while *B'//BP* paramutates *B-I* as efficiently as *B'*.

The involvement of repeated sequences has been proposed for other cases of gene inactivation in plants, but counter-examples make it clear that complex repeated structures of the type seen at *r* are not required. Some dominant alleles of the *nivea* gene in snapdragon act *in trans* to suppress the accumulation of RNA from *Niv+*. These dominant alleles are mutant derivatives of *Niv+*, and each has a duplication of part of the gene, including promoter sequences. Bollman et al. (1991) suggest that this complex structure causes *trans*-inactivation. However, despite a positive correlation with transgene copy number and the inactive state (Linn et al. 1990; Scheid et al. 1991), a number of cases of inactivation involve a single integrated copy of a transgene (Linn et al. 1990; van der Krol et al. 1990; Smith et al. 1990; Seymour et al. 1993; Meyer et al. 1993).

4 Effect of Paramutation on Gene Structure

Since *b* paramutation is associated with changes in transcription, efforts were made to identify changes in sequence or structure of putative regulatory regions of *B'* and *B-I* (PATTERSON et al. 1993b). No DNA rearrangements were detected in a region from −11 000 bp to +13 000 bp (relative to the transcription start at +1), as measured by the patterns of bands seen on Southern blots. *B'* and *B-I* clones were sequenced from −1487 to +730, and were identical. Cutting of *B'* and *B-I* with a large number of methylation-sensitive restriction enzymes showed that the methylation pattern of the two alleles was indistinguishable, from 6 kb upstream of the transcription start, through the 4-kb *b* coding region, to 2 kb downstream of the poly-A addition site.

The lack of correlation of DNA methylation with *b* expression might be considered surprising, since related phenomena do show a correlation. *r* paramutation is accompanied by substantial increases in methylation of the *R-r* gene complex, which contains *R(S)* (M. Alleman and J. Kermicle, personal communication; DOONER et al. 1991). Several examples of transgene interactions that lead to gene silencing are also correlated with increases in methylation (MATZKE and MATZKE 1991; MEYER et al. 1993; ASSAAD et al. 1993). In the *Spm* and *Ac* transposable element systems in maize, an inactive element that is not making transposase can spontaneously undergo a change that results in its activation, and this change is correlated with hypomethylation of sites in CpG islands of the elements. An active transposable element can increase the likelihood that an inactivate element is reactivated, and this reactivation is also accompanied by hypomethylation (reviewed in FEDOROFF and CHANDLER 1994). In fungi, interactions between duplicated genes can lead to heritable changes in DNA methylation (Singer and Selker, this volume). Parental imprinting in mammals is correlated with DNA methylation (reviewed in SURANI 1993), as is X-chromosome inactivation in eutherian mammals (LOCK et al. 1987). It is worth emphasizing that, unlike the examples above, *b, r,* and *pl* paramutation do not result in complete silencing. Pigment accumulation is still seen, and in the case of *B'* and *Pl'-mah*, transcription is still readily detectable.

The difference between *b* and these other phenomena can be explained in two ways. First, it may be that *b* paramutation is not correlated with DNA methylation; and second, it may be that subtle changes in methylation exist, but went undetected in our analysis. Given these two possibilities, it is instructive to examine the exact nature of the correlation between DNA methylation and these related gene inactivation phenomena. In the case of *b* paramutation, it was estimated that ~20% of the cytosines that are contained in canonical methylatable sequences were tested, in the region from ~−3500 to ~+5500 (relative to the transcription start at +1). This survey included two CpG island in the vicinity of the transcription start site (PATTERSON et al. 1993b). In other cases in which heritable changes in gene expression are correlated with differences in methylation, changes in DNA methylation were observed by surveying a much smaller number of sites (Fig. 3). This comparison emphasizes that the correlation of DNA

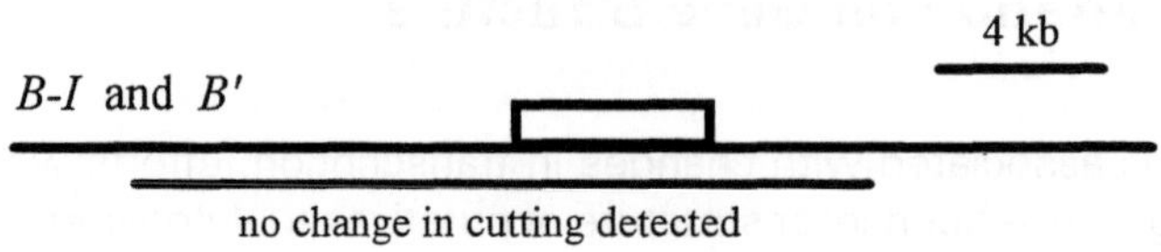

a

	# sites changed[a]	# tested	% changed	reference
Paramutation				
b	0	43[b]	0	PATTERSON et al. 1993b
r	---[c]	---	many	M ALLEMAN and J KERMICLE personal communication
Transgene interactions				
Petunia	5	5	100	MEYER et al. 1993
Tobacco	2	9	22	MATZKE and MATZKE 1991
Transposable element interactions				
Spm	6	16	38	BANKS et al. 1988
Ac	6	8	75	CHOMET et al. 1987 SCHWARTZ 1989
Imprinted genes in mice				
H19	11	23	48	FERGUSON-SMITH et al. 1993 BARTHOLEMEI et al. 1993
Igf2r[d]	63	64	98	STOGER et al. 1993
Igf2	4	25	16	SASAKI et al. 1992 BRANDEIS et al. 1993

b

Fig. 3 a, b. Methylation comparison. **a** A schematic map of *b. B'* and *B-I* are represented by a single map, as no differences in DNA structure or sequence have been seen. The *box* represents the transcribed region of the gene. The *line under the map* indicates the region of *B'* and *B-I* tested for differences in DNA methylation. **b** Comparison of *b* to other examples of meiotically heritable alterations in gene expression. *Notes: a,* "# sites changed" refers to the number of restriction sites of methylation-sensitive restriction enzymes of which cutting does not occur when DNA is isolated from individuals with the less active gene state. *b,* The 43 sites shown refer to those mapped from ~−3500 to ~+1000 (relative to the transcription start). A number of additional sites were tested across the rest of the coding region and through two kilobase pairs of 3' flanking DNA. The number of sites tested in this region is not known precisely, but it was estimated that ~20% of the cytosines that are contained in canonical methylatable sequences were tested throughout this region (PATTERSON et al. 1993b). *c,* Restriction sites in *R(S)* and *R(S)'* were difficult to map because *r* is part of a gene complex, and the details of its structure are currently under study. It is clear that the changes in DNA methylation are extensive, and a majority of sites tested are hypermethylated in *R(S)'* relative to *R(S). d,* The 64 sites shown here represent sites from only a subset of the >100 kb pair transcribed region of this gene (see STOGER et al. 1993 for details)

methylation and gene expression is different for *b* than for any of the other examples; when changes in DNA methylation are seen, the changes are extensive enough to be obvious. However, methylation changes during *b* paramutation that are unusually distant from the gene or alter only an unusually small number of cytosine residues cannot be ruled out.

The causal relationship between gene expression and cytosine methylation is not clear. In some cases DNA methylation happens after, not before, changes in gene expression (Lock et al. 1987; Enver et al. 1988). Hence, in these cases, methylation does not establish gene silencing but may play a role in maintaining a silent state. In some cases, it is clear that DNA methylation is required for the maintenance of a silent state. In mice that have a defective DNA methyltransferase gene, imprinted genes are neither hypermethylated nor imprinted (Li et al. 1993). Transfection experiments showed that DNA from the hypermethylated, inactive X-chromosome of humans or mice is incapable of being expressed (Venolia and Gartler 1983; Liskay and Evans 1980). Finally DNA methylation is not always associated with gene silencing. In marsupials, and in mouse extraembryonic tissues, X-inactivation is not correlated with cytosine modification (Kaslow and Migeon 1987; Kratzer et al. 1983). Transcriptionally silent states are maintained in yeast and *Drosophila*, which have no detectable cytosine methylation (see Sect. 6). Riggs and Pfeifer (1992) observed that late DNA replication is more often correlated with gene inactivity. Such silencing phenomena as X-chromosome inactivation, silencing of genes at the mating type loci in yeast, and *Drosophila* position effect variegation are associated with late DNA replication. Mutational analysis has indicated that DNA replication is required for the establishment of gene silencing in yeast, but the mechanistic role of replication is uncertain (Laurenson and Rine 1992).

5 Models for Paramutation

Models should explain how the alleles interact, and also, how that interaction causes heritable alterations. Genetic models propose that an inducing allele such as *B′* transfers a sequence element by gene conversion to a sensitive allele, or induces a sequence change by other means. No physical evidence for a sequence change between *B-I* and *B′* has been seen; however, as discussed in Sect. 4, the sequence change may be subtle or physically distant from the *b* locus. *B-I* frequently changes to *B′* in a *B-I/B-I* plant; therefore, a sequence change responsible for paramutation must be capable of occurring spontaneously at a high frequency. Several diseases in humans result from an increase in copy number of nucleotide triplet repeats (reviewed in Warren and Nelson 1993); a similar sort of increase in a repeated sequence element could be responsible for paramutation. However, it is not clear how an allelic interaction would lead to directed increases in the number of repeats. There are a few examples of unidirectional gene conversion events that are nearly as regular as paramutation, such as the transfer of infectious intervening sequences (Gimble and Thorner 1992) and mating type switching in yeast (reviewed in Haber 1983).

A second class of models proposes epigenetic changes in paramutated genes. In these models, allelic interactions lead not to changes in DNA sequence, but to epigenetic changes, such as changes in chromatin structure or DNA methylation. The alleles might interact in two ways:

1. Homology might be recognized by DNA-DNA interactions, perhaps similar to those which occur during homologous recombination (reviewed in Dooner et al. 1991). In *Neurospora* and yeast, and in *Drosophila*, the machinery for identifying homologous sequences in mitotically dividing cells can be quite efficient. Maize has a much larger genome, and there is no clear evidence on the question of whether homologous sequences can interact in somatic cells. In the mouse, which has a comparably large and complex genome, homologous recombination can occur in somatic cells, as evidenced by gene disruption technology, but this process is not highly efficient (Thomas and Capecchi 1987).

2. Proteins bound to a regulatory sequence on one allele might interact with proteins bound to the homologous allele. This interaction is exactly analogous to models suggested for transvection in *Drosophila*. In transvection, a transcriptional regulatory sequence from one allele acts to increase or decrease expression from an allele on the homologous chromosome (Henikoff and Dreesen 1989; Geyer et al 1990).

Two models have been proposed for how allelic interactions might result in heritable epigenetic changes in gene expression:

1. The interaction might result in differential cytosine methylation. For example, the unusual structure that resulted from the pairing of alleles might affect the accessibility of the genes or associated regulatory sequences to methylase (see Sect. 4 for discussion of methylation as it relates to *b* and *r* paramutation; for review, see Matzke and Matzke 1993; Jorgensen 1993).

2. The interaction might result in a change in chromatin structure. Brink (1964) and Patterson et al. (1993b) have proposed models for paramutation that involve such a change.

One attractive model that unifies observations in animals and fungi is that late DNA replication and changes in chromatin structure reinforce each other, and lead to gene silencing (Riggs and Pfeifer 1992). In some cases cytosine methylation is proposed to play a role in maintaining either late replication or chromatin structure. Paramutation is a meiotically heritable change in a gene, and changes in chromatin structure are generally thought to be erased during gametogenesis and meiosis. There is some resistance to the idea that changes in chromatin structure might be inherited; however, it has been clear for decades that epigenetic changes in plant genes that are acquired during development can be inherited (see Jorgensen 1993 for review). Recent work on imprinting in mammals and on heritable transgene inactivation in plants has led to a renewed interest in this problem. It is widely accepted that changes in chromatin structure are mitotically heritable (see below). Similar changes might in special circumstances be heritable through gametogenesis and meiosis. In the following section, we discuss the somatic nature of paramutation, and compare it to somatic changes in chromatin structure that occur in animals and yeast.

6 Comparison of Paramutation to Somatic Epigenetic Changes in Other Organisms

In thinking about a mechanism for paramutation, the question of when in development a sensitive allele is altered is important. Paramutation may result from an interaction in somatic tissues, which is transmitted because, in plants, the germline in derived very late in development from the same cell lineages that give rise to the soma (see below for further discussion).

The available evidence suggests that paramutation occurs in somatic tissues. Spontaneous changes of *B-I* to *B'* can be seen as somatic sectors in *B-I/B-I* plants (Coe 1966). Another indication of the somatic nature of paramutation is shown in Fig. 1a and b. In *B'/B-I* and *Pl'-mah/Pl-Rh* heterozygotes, the pigmenting potential of the sensitive alleles *B-I* and *Pl-Rh* are drastically reduced in somatic tissues. The reduction of pigment in the somatic tissues of these F1 plants is equivalent to the heritable reduction seen in the F2. While *R-st* has a very weak effect on the phenotype of *R(S)* in the F1, other genetic evidence indicates that *r* paramutation also occurs in somatic tissues (Sastry et al. 1965). The most direct evidence for the somatic nature of paramutation is seen in plant species other than maize (discussed in Sastry et al. 1965 and Brink 1973). In these cases, paramutation is clearly progressive in the body of the plant; that is, branches that develop late in the life cycle show more severe effects somatically, and flowers on the more severely affected branches produce more severely affected progeny. Thus, in these plants, F1 paramutation is apparently identical to heritable paramutation. Current evidence is also consistent with *b* paramutation being a somatic event. Coe suggested that *b* paramutation, unlike other examples of paramutation, is perhaps a meiotic event, but this suggestion has been reevaluated in light of recent work (Coe 1966; Patterson 1993; Patterson and Chandler, manuscript in preparation).

Work in plants, vertebrate and invertebrate animals, and fungi has demonstrated that stable, mitotically heritable epigenetic changes occur during the normal development and growth of eukaryotic organisms. In Sect. 4, we discussed the examples of imprinting and X-chromosome inactivation in mammals. Experiments in *Drosophila* and yeast, two organisms that have no cytosine methylation, demonstrate the existence of mitotically heritable transcription states (Laurenson and Rine 1992; Paro 1990). For example, the *Drosophila* DNA binding protein hunchback negatively regulates *ultrabithorax* transcription (Qian et al. 1991). However, suppression of *ultrabithorax* is observed much later in development, after the hunchback protein is gone (Tautz 1988; Bender et al. 1987). Genes of the silent mating type loci in yeast can be transcriptionally active or inactive, and each state is mitotically heritable. Switches from one transcription state to the other do occur (Pillus and Rine 1989; Mahoney et al. 1991). Mutations in chromatin proteins that destabilize these inactive states have led to the hypothesis that changes in chromatin structure are responsible (Paro 1990; Laurenson and Rine 1992).

In plants, as in animals, most developmentally programmed epigenetic changes are not transmitted to progeny, but rather, are reset during gametogenesis. However, there are a myriad of exceptions to this rule in plants, among them paramutation (see JORGENSEN 1993 for review). Other examples include the heritable *trans*-inactivation of transgenes (MEYER et al. 1993; GORING et al. 1991; MATZKE and MATZKE 1991; reviewed in MATZKE and MATZKE 1993) and parental imprinting of *r* and other genes in plants (KERMICLE 1970b; KERMICLE and ALLEMAN 1990). The phenomenon of transposable element presetting is in some ways similar to paramutation. An examples of this phenomenon is seen when the expression of the *a1-m2* allele, which contains a defective *Spm* element (*dSpm*), is induced by the presence of an unlinked, active *Spm* element (MCCLINTOCK 1964). This induction can be seen in the next generation, even when the active element has segregated away, and can occasionally persist for two more generations (MCCLINTOCK 1965). A meiotically heritable change in the expression of the transposable element-containing *niv-53* allele of the *nivea* gene in *Antirhinnum majus* has been proposed to be caused by unlinked transposable elements (KREBBERS et al. 1987). This interaction, originally described as an example of paramutation, is perhaps more like presetting. Other transposable element regulatory events also show this type of meiotic inheritance (reviewed in FEDOROFF and CHANDLER 1994). The maize elements *Mu, Ac,* and *Spm* go through phases of activity and inactivity, and these phases are associated with hypomethylation and hypermethylation respectively. The phase of an element has been shown to be affected by developmental events and in the case of *Spm*, regulatory signals from other transposable elements, and the changes that occur are often passed on to progeny. In a number of cases, the expression of a maize gene with a *Mu* or *Spm* insertion is negatively regulated such that cycles of transposable element activity and inactivity lead to cycles of gene inactivity and activity. In all of these transposable element regulatory events, epigenetic change during development is passed on to progeny. In a final example, MIKULA (1967) has shown that the expression of the *r* gene in maize is affected by day-length conditions in the previous generation. Perhaps the only known animal genes in which developmentally programmed epigenetic changes are transmitted are genes that are differentially imprinted in the male and female germline (reviewed in SURANI 1993).

The examples given above indicate that meiotic inheritance of epigenetic changes is much more commonly observed in plants than in animals. Why might plants and animals differ in heritability of epigenetic changes? As stated previously, differences in plant and animal development can provide an explanation. The germline and the soma of plants are not separate lineages until late in development. In contrast, in most animals, these two lineages separate very early in development. In *Caenorhabditis elegans*, the germline becomes separate from soma at the fourth cell division after fertilization (SULSTON 1988). The epigenetic changes that are necessary for somatic cell lineages are typically isolated from the germline, and not transmitted to progeny. In maize, a population of stem cells (the meristem) produces most plant organs before separation of germline and somatic lineages. Sectors of tissue that are less than 1% of the male inflorescence can include both somatic cells and gametes (DAWE and FREELING 1990). Thus, if a gene undergoes an epigenetic change during development, the cell in which the change occurs may give rise to both soma and germline. Changes such as paramutation may be more likely to be inherited in plants, not because the physical nature of gene alteration is necessarily different, but because plant and animal development are different.

It is possible that the germline in plants is less efficient at erasing epigenetic information, or that the mechanism of imprinting epigenetic changes is fundamentally different in plants. It is difficult to evaluate these ideas without more information on the molecular basis of epigenetic changes.

The above speculations about the mechanism of inheritance of epigenetic changes in plants raise a related question: Does this form of inheritance serve any purpose in plants? It may be that this type of inheritance is a mistake, a failure to appropriately erase epigenetic changes that occur during development (BRINK 1964). Alternatively, the inheritance of these changes may serve to program appropriate gene expression in offspring, and thus serve an adaptive purpose.

7 How Common Is Paramutation?

To identify paramutation, both sensitive alleles such as *B′* and responsive alleles such as *B-I* must exist for a locus. If one considers *b* paramutation as a paradigm, populations with both types of alleles would be rare, because *B′* changes *B-I* into *B′*, and *B′* is very stable (Sect. 2.1). Possibly, genes that have the potential to undergo paramutation are more common than is recognized, but the observation of paramutation requires the identification of a rare, unstable allele like *B-I*, that is only in a population transiently. The opportunity for identifying such rare mutants may have been extensive, as human selection for new color variants of maize has been ongoing, perhaps for thousands of years.

The anthocyanin pathway may be unusually well suited to identifying paramutation and related phenomena. Interesting mutations can be recovered and easily propagated because variants can be easily identified by visual inspection and anthocyanins are not essential for plant growth. The pathway is very sensitive to expression of the regulatory genes; the tenfold difference in *b* transcription between *B′/B′* and *B-I/B-I* plants produces a dramatic change in phenotype. Even *B′/B-I* and *B′/B′*, which have a barely detectable difference in *b* RNA levels, have slight, but detectable differences in phenotype (COE 1966; PATTERSON et al. 1993b; Patterson and Chandler, manuscript in preparation). Thus, even a subtle change in an anthocyanin regulatory gene can be easily identified.

8 Concluding Remarks

Paramutation is an allelic interaction that has been shown to result in a mitotically and meiotically heritable change in transcription state. It is clear that meiotically heritable changes in gene expression are observed more often in plants than in animals, but we suggest that this particular difference in phenomenology may be due to a difference in animal and plant development rather than a difference in mechanism.

Acknowledgements. We thank Ken Kubo, Jay Hollick, Dave Hawthorne, Manuel Sainz, Eric Selker, Bryan Pickett, Wim van Heeckeren, and Terry Shroyer for critical comments on the manuscript. We are grateful to Mary Alleman, Jerry Kermicle, William Eggleston, Ken Kubo, Terry Shroyer, and Jay Hollick for sharing unpublished data. Work supported by a National Institutes of Health grant (GM35791) to V.L.C.

References

Ashman RB (1970) The compound structure of the R^{st} allele in maize. Genetics 64:239-245

Assaad FF, Tucker KL, Signer ER (1993) Epigenetic repeat-induced gene silencing (RIGS) in Arabidopsis. Plant Mol Biol 22: 1067–1087

Banks JA, Masson P, Fedoroff N (1988) Molecular mechanisms in the developmental regulation of the maize Suppressor-mutator transposable element. Genes Dev 2: 1364–1380

Bartolomei MS, Webber AL, Brunkow ME, Tilghman SM (1993) Epigenetic mechanisms underlying the imprinting of the mouse H19 gene. Genes Dev 7: 1663–1673

Bender M, Turner FR, Kaufman TC (1987) A developmental genetic analysis of the gene Regulator of postbithorax in Drosophila melanogaster. Dev Biol 119: 418–437

Bodeau JP, Walbot VW (1992) Regulated transcription of the maize Bronze-2 promoter in electroporated protoplasts requires the C1 and R gene products. Mol Gen Genet 233: 379–387

Bollmann J, Carpenter R, Coen ES (1991) Allelic interactions at the nivea locus of Antirrhinum. Plant Cell 3: 1327–1336

Brandeis M, Kafri T, Ariel M, Chaillet JR, McCarrey J, Razin A, Cedar H (1993) The ontogeny of allele-specific methylation associated with imprinted genes in the mouse. EMBO J 12: 3669–3677

Brink RA (1958) Basis of a genetic change which invariably occurs in certain maize heterozygotes. Science 127: 1182–1183

Brink RA (1964) Genetic repression of R action in maize. In: Locke M (ed) The role of chromosomes in development. The 23rd symposium of the Society for the Study of Development and Growth. Academic, New York, pp 183–230

Brink RA (1973) Paramutation. Annu Rev Genet 7: 129–152

Brink RA, Styles ED, Axtell JD (1968) Paramutation: directed genetic change. Science 159: 161–170

Brink RA, Kermicle JL, Ziebur NK (1970) Derepression in the female gametophyte in relation to paramutant R expression in maize endosperms, embryos and seedlings. Genetic 66: 87–96

Brown DF (1966) Paramutability of R^g and r^r mutant genes derived from an R^r allele in maize. Genetics 54: 899–910

Brown DF, Brink RA (1960) Paramutagenic action of paramutant R^r and R^g alleles in maize. Genetics 45:1313-1316

Brusslan JA, Karlin -Neumann GA, Huang L, Tobin EM (1993) An arabidopsis mutant with a reduced level of cab 140 RNA is a result of cosuppression. Plant Cell 5: 667–677

Chomet PS, Wessler S, Dellaporta S (1987) Inactivation of the maize transposable element Activator (Ac) is associated with its DNA modification. EMBO J 6: 295–302

Clark JK, Chandler VL, Neuffer MG (1990) Characterization of two Ds mutants of B-Peru. Maize Genet Coop Newsl 64: 59–60

Cocciolone SM, Cone KC (1993) Pl-bh, an anthocyanin regulatory gene of maize that leads to variegated pigmentation. Genetics 135: 575–588

Coe EH (1966) The properties, origin and mechanism of conversion-type inheritance at the B locus in maize. Genetics 53: 1035–1063

Coe EH (1979) Specification of the anthocyanin biosynthetic function by b and r in maize. Maydica 24: 49–58

Coe EH, Neuffer MG, Hoisington DA (1988) The genetics of corn. In: Sprague G, Dudley JW (eds) Corn and corn improvement, American Society of Agronomy, Madison, Wisconsin, pp 83–258

Cone KC, Cocciolone SM, Burr FA, Burr B (1993) Maize anthocyanin regulatory gene pl is a duplicate of c1 that functions in the plant. Plant Cell 5: 1795–1805

Dawe RK, Freeling M (1990) Clonal analysis of the cell lineages in the male flower of maize. Dev Biol 142: 233–245

de Carvalho F, Gheysen G, Kushnir, S, van Montague M, Inze D, Castresana C (1992) Suppression of beta-1,3-glucanase transgene expression in homozygous plants. EMBO J 11: 2595–2602

Dooner HK, Kermicle JL (1971) Structure of the R^r tandem duplication in maize. Genetics 67: 427–436

Dooner HK, Robbins TP, Jorgenson RA (1991) Genetic and developmental control of anthocyanin biosynthesis. Annu Rev Genet 25: 173–199

Elkind Y, Edwards R, Mavandad M, Hedrick SA, Ribak O, Dixon RA, Lamb CJ (1990) Abnormal plant development and down-regulation of phenylpropanoid biosynthesis in transgenic tobacco containing a heterologous phenylalanine ammonia-lyase gene. Proc Natl Acad Sci USA 87: 9057–9061

Enver T, Zhang J, Papayannopoulou T, Stamatoyannopoulos G (1988) DNA methylation: a secondary event in globin gene switching? Genes Dev 2: 698–706

Fedoroff NV, Chandler VL (1994) Inactivation of maize transposable elements. In Paszkowski J (ed) Homologous recombination in plants. Kluwer Academic, Dordrecht, The Netherlands, (in press)

Ferguson-Smith AC, Sasaki H, Cattanach BM, Surani MA (1993) Parental-origin-specific epigenetic modification of the mouse H19 gene. Nature 362: 751–755

Geyer PK, Green MM, Corces VG (1990) Tissue-specific transcriptional enhancers may act in trans on the gene located in the homologous chromosome: the molecular basis of transvection in Drosophila. EMBO J 9: 2247–2256

Gimble FS, Thorner J (1992) Homing of a DNA endonuclease gene by meiotic gene conversion in Saccharomyces cerevisiae. Nature 357: 301–306

Goff SA, Klein TM, Roth BA, Fromm ME, Cone KC, Radicella JP, Chandler VL (1990) Transactivation of anthocyanin biosynthetic genes following transfer of b regulatory genes in maize tissues EMBO J 9: 2517–2522

Goff SA, Cone KC, Chandler VL (1992) Functional analysis of the transcriptional activator encoded by the maize B gene: evidence for a direct functional interaction between two classes of regulatory proteins. Genes Dev 6: 864–875

Goring DR, Thomson L, Rothstein SJ (1991) Transformation of a partial nopaline synthase gene into tobacco suppresses the expression of a resident wild-type gene. Proc Natl Acad Sci USA 88: 1770–1774

Grierson D, Fray RG, Hamilton AJ,Smith CJS, Watson CF (1991) Does co-suppression of sense genes in transgenic plants involve antisense RNA? Trends Biotechnol 9: 122–123

Haber JE (1983) Mating type genes of Saccharomyces cerevisiae. In: Shapiro J (ed) Mobile genetic elements, Academic, New York, pp 560–621

Henikoff S, Dreesen TD (1989) Trans-inactivation of the Drosophila brown gene: evidence for transcriptional repression and somatic pairing dependence. Proc Natl Acad Sci USA 86: 6704–6708

Jorgensen R (1993) The germinal inheritance of epigenetic information in plants. Philos Trans R Soc Lond [Biol] 339: 173–181

Kaslow DC, Migeon B (1987) DNA methylation stabilizes X-chromosome inactivation in eutherians but not in marsupials: evidence for multistep maintenance of mammalian X-dosage compensation. Proc Natl Acad Sci USA 84: 6210–6214

Kermicle JL (1970a) Somatic and germinal instability of R-stippled, an aleurone spotting factor in maize. Genetics 64: 247–258

Kermicle JL (1970b) Dependence of the r-mottled aleurone phenotype in maize on mode of sexual transmission. Genetics 66: 69–85

Kermicle JL (1974) Organization of paramutational components of the R locus in maize. Brookhaven Symp Biol 25: 262–280

Kermicle JL, Alleman M (1990) Gametic imprinting in maize in relation to the angiosperm life cycle. Development [Suppl] 9–14

Kratzer PG, Chapman VM, Lambert H, Evans RE, Liskay RM (1983) Differences in the DNA of the inactive X-chromosomes of fetal and extraembryonic tissues of mice. Cell 33: 37–42

Krebbers E, Hehl R, Piotrowiak R, Lonnig W, Sommer H, Saedler H (1987) Molecular analysis of paramutant plant of Antirhinnum majus and the involvement of transposable elements. Mol Gen Genet 209: 499–507

Laurenson P, Rine J (1992) Silencers, silencing and heritable transcriptional states. Microbiol Rev 56: 543–560
Li E, Beard C, Jaenisch R (1993) Role for DNA methylation in genomic imprinting. Nature 366: 362–365
Linn F, Heidmann I, Saedler H, Meyer P (1990) Epigenetic changes in the expression of the maize A1 gene in Petunia hybrida: Role of numbers of integrated gene copies and state of methylation. Mol Gen Genet 222: 329–336
Liskay RM, Evans RJ (1980) Inactive X-chromosome DNA does not function in DNA-mediated cell transformation for the hypoxanthine phosphoribosyltransferase gene. Proc Natl Acad Sci USA 77: 4895–4898
Lock LF, Takagi N, Martin G (1987) Methylation of the Hprt gene on the inactive X occurs after chromosome inactivation. Cell 48: 39–46
Ludwig SR, Habera LF, Dellaporta SL, Wessler SR (1989) Lc, a member of the maize R gene family responsible for tissue-specific anthocyanin production, encodes a protein similar to transcriptional activators and contains the myc-homology region. Proc Natl Acad Sci USA 86: 7092–7096
Mahoney DJ, Marquardt R, Shei G, Rose AB, Broach JR (1991) Mutations in the HML E silencer of Saccharomyces cerevisiae yield metastable inheritance of transcriptional repression. Genes Dev 5: 605–615
Matzke MA, Matzke AJM (1991) Differential inactivation and methylation of a transgene in plants by two suppressor loci containing homologous sequeneces. Plant Mol Biol 16: 821–830
Matzke M, Matzke AJM (1993) Genomic imprinting in plants: parental effects and trans-inactivation phenomena. Annu Rev Plant Physiol Plant Mol Biol 44: 53–76
McClintock B (1964) Aspects of gene regulation in maize. Carnegie Inst Washington Year Book 63: 592–602 (Reprinted in: The Discovery and Characterization of Transposable Elements, 1987, Garland, New York, pp 452–465
McClintock B (1965) Components of action of the regulators Spm and Ac. Carnegie Inst of Washington Year Book 64: 527–536 (Reprinted in: The Discovery and Characterization of Transposable Elements, 1987, Garland, New York, pp 466–477)
McWhirter KS, Brink RA (1963) Paramutation in maize during endosperm development. Genetics 48: 189–203
Meyer P, Heidmann I, Niedenhoff I (1993) Differences in DNA methylation are associated with a paramutation phenomenon in transgenic petunia. Plant J 4: 89–100
Mikula BC (1967) Heritable changes in r-locus expression in maize in response to environment. Genetics 56: 733–742
Mol J, van Blokland R, Kooter J (1991) More about co-suppression. Trends Biotech 9: 182–183
Paro R (1990) Imprinting a determined state into the chromatin of Drosophila. Trends Genet 6: 416–421
Patterson GI (1993) Paramutation: a directed, meiotically heritable change in gene expression. PhD thesis, University of Oregon library, Eugene, Oregon
Patterson GI, Harris LJ, Walbot V, Chandler VL (1991) Genetic analysis of B-Peru, a regulatory gene in maize. Genetics 126: 205–220
Patterson GI, Kaszas E, Cone KC, Coe EC, Chandler VL (1993a) The *pl* gene undergoes paramutation. Maize Genet Coop Newsl 67: 42–43
Patterson GI, Thorpe CJ, Chandler VL (1993b) Paramutation an allelic ineteraction, is associated with a stable and heritable reduction of transcription of the maize b regulatory gene. Genetics 135: 881–894
Paz-Ares J, Ghosal D, Weinand U, Peterson PA, Saedler H (1987) The regulatory c1 locus of Zea mays encodes a protein with homology to myb proto-oncogene products and with structural similarities to transcriptional activators. EMBO J 6: 3553–3558
Pillus L, Rine J (1989) Epigenetic inheritance of transcriptional states in S. cerevisiae. Cell 59: 637–647
Qian S, Capovilla M, Pirrotta V (1991) The bx region enhancer, a distant cis-control element of the Drosophila Ubx gene and its regulation by hunchback and other segmentation genes. EMBO J 10: 1415–1425
Radicella JP, Turks D, Chandler VL (1991) Cloning and nucleotide sequence of a cDNA encoding B-Peru, a regulatory protein of the anthocyanin pathway in maize. Plant Mol Biol 17: 127–130
Radicella JP, Brown D, Tolar LA, Chandler VL (1992) Allelic diversity of the maize b regulatory gene: different leader and promoter sequences of two b alleles determine distinct tissue specificities of anthocyanin production. Genes Dev 6: 2152–2164
Riggs AD, Pfeiffer GP (1992) X-chromosome inactivation and cell memory. Trends Genet 8: 169–174
Robbins TP, Walker EL, Kermicle JL, Alleman M, Dellaporta SL (1991) Meiotic instability of the R-r complex arising from displaced intragenic exchange and intrachromosomal rearrangement. Genetics 129: 271–283

Roth BA, Goff SA, Klein TM, Fromm ME (1991) C1- and R-dependent expression of the maize Bz1 gene requires sequences with homology to mammalian myb and myc binding sites. Plant Cell 3: 317–325
Rothstein SJ, DiMaio J, Strand M, Rice D (1987) Stable and heritable inhibition of the expression of nopaline synthase in tobacco expressing antisense RNA. Proc Natl Acad Sci USA 84: 8439–8443
Sasaki H, Jones PA, Chaillet JR, Ferguson-Smith AC, Barton SC, Reik W, Surani MA (1992) Parental imprinting: potentially active chromatin of the repressed maternal allele of the mouse insulin-like growth factor II (Igf2) gene. Genes Dev 6: 1843–1856
Sastry GRK, Cooper HB, Brink RA (1965) Paramutation and somatic mosaicism in maize. Genetics 52: 407–424
Scheid OM, Paszkowski J, Potrykus I (1991) Reversible inactivation of a transgene in Arabidopsis thaliana. Mol Gen Genet 228: 104–112
Schwartz D (1989) Gene-controlled cytosine demethylation in the promoter region of the Ac transposable element in maize. Proc Natl Acad Sci USA 86: 2789–2793
Seymour GB, Fray RG, Hill P, Tucker GA (1993) Down-regulation of two non-homologous endogenous tomato genes with a single chimaeric sense gene construct. Plant Mol Biol 23: 1–9
Smith CJS, Watson CF, Bird CR, Ray J, Schuch W, Grierson D (1990) Expression of a truncated tomato polygalacturonase gene inhibits expression of the endogenous gene in transgenic plants. Mol Gen Genet 224: 477–481
Stadler LJ, Nuffer MG (1953) Problems of gene structure. II. Separation of R^r elements (S) and (P) by unequal crossing over. Science 117: 471–472
Stoger R, Kubicka P, Liu C, Kafri T, Razin A, Cedar H, Barlow Dep (1993) Maternal-specific methylation of the imprinted mouse Igf2r locus identifies the expressed locus as carrying the imprinting signal. Cell 73: 61–71
Styles ED, Brink RA (1969) The metastable nature of paramutable R alleles in maize. IV. Parallel enhancement of R action in heterozygotes with r and in hemixygotes. Genetics 61: 801–811
Styles ED, Ceska O, Seah K (1973) Developmental differences in action of r and b alleles in maize. Can J Genet Cytol 15: 59–72
Sulston J (1988) Cell lineage. In: Wood WB (ed) The nematode caenorhabditis elegans. Cold Spring Harbor Laboratory, Cold Spring Harbor, New York, pp 123–155
Surani MA (1993) Silence of the genes. Nature 366: 302–303
Tautz D (1988) Regulation of the Drosophila segmentation gene hunchback by 2 maternal morphogenetic centers. Nature 332: 281–284
Thomas KR, Capecchi MR (1987) Site-directed mutagenesis by gene-targeting in mouse embryo-derived stem cells. Cell 51: 503–512
van der Krol AR, Mol LA, Beld M, Mol JNM, Stuitje AR (1990) Flavanoid genes in petunia: addition of a limited number of gene copies may lead to a suppression of gene expression. Plant Cell 2: 291–299
van der Walt WJ, Brink RA (1969) Geographic distribution of paramutable and paramutagenic R alleles in maize. Genetics 61: 677–695
Venolia L, Gartler SM (1983) Comparison of transformation efficiency of human active and inactive X-chromosome DNA. Nature 302: 82–83
Warren ST, Nelson DL (1993) Trinucleotide repeat expansions in neurological disease. Curr Opin Neurobiol 43: 752–759

DNA Methylation and Activity of the Maize *Spm* Transposable Element

N.V. Fedoroff

Carnegie Institution of Washington, Department of Embryology, 115 West University Parkway, Baltimore, MD 21210, USA

1 Introduction

Maize transposons were the first plant genes known to undergo reversible heritable inactivation. Early in the study of the *Suppressor-mutator* (*Spm*) transposable element McClintock recognized that certain isolates of the element either cycled between inactive and active phases during development or underwent an inactivation event of longer duration and sufficient stability to be heritable, but which was nonetheless occasionally reversed (McClintock 1957, 1958). In subsequent studies, McClintock developed a deeper understanding of the ways in which the *Spm* element alternated between active and inactive phases (McClintock 1959, 1961, 1962, 1971). She later reported that the *Activator* element is also subject to a similar type of reversible inactivation, although the she did not study the *Activator* element's inactivation mechanism in as great detail as that of *Spm* (McClintock 1964, 1965a,b).

An indication that transposon inactivity might be associated with DNA methylation emerged early in the analysis of maize genomic sequences homologous to the *Ac* element, first cloned from the *waxy* locus (Fedoroff et al. 1983). Despite the presence of multiple sequences resembling the element in the genome, only the single active copy of the *Ac* element located at the *wx* locus could be cleaved by enzymes whose ability to cut DNA was inhibited by the methylation of C residues within their cleavage sites. This observation subsequently led not only to extensive analysis of transposable element methylation, but has proved useful for cloning genes known to be marked by the insertion of an active element (Cone et al. 1988). Both the *Ac* element and the more recently discovered *Mu* element families are subject to genetic inactivation associated with methylation of element sequences (Chandler and Walbot 1986; Bennetzen 1987). While the inactivation mechanisms that affect these element families have not yet been investigated in as great detail as that of the *Spm* element, the subject of the present review, it is already clear that there are strong parallels.

One of the major difficulties in understanding transposable element inactivation is that transposition can cease for any one of several reasons. The first is that the transposon cannot express element-encoded gene products required for its own transposition. The second is that an element mobilizing a transposition-defective element under observation has been inactivated. And the third is that the element has been altered so that it fails to transpose even when supplied *in trans* with all of the proteins necessary for transposition. In the present review, I will focus primarily on the genetics and molecular biology of a mechanism that controls the transcriptional activity of the *Spm* element. The common ability of maize transposable elements to exist in a heritably, but reversibly inactive form poses a number of questions not only about the molecular mechanisms of inactivation and reactivation, but also about the importance to the evolutionary survival of transposable elements of their capacity to persist in the genome in a silent form.

2 The Structure of the *Spm* Element Family

The *Spm* element family contains two types of transposition-competent element, as well as transposition-defective elements. The intact *Spm* element is 8.3 kb in length and virtually identical in length and sequence with the independently identified *En* element (PEREIRA et al. 1986; MASSON et al. 1987; Fig. 1). There are many internally deleted derivatives that either transpose at a reduced frequency or are completely transposition-defective. There are several derivatives of the *Spm* and *En* elements, termed *Spm-w* and *En-low*, that transpose at a reduced

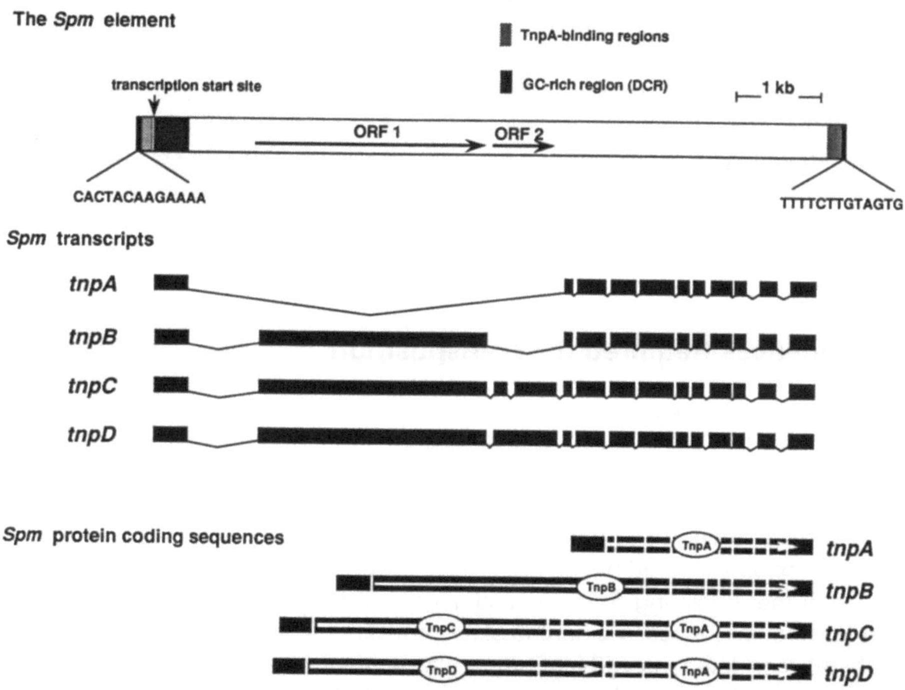

Fig. 1. A diagrammatic representation of the *Spm* element's structure, transcripts and protein-coding sequences. The *open box* at the *top* of the diagram represents the 8.3-kb *Spm* element and the *filled ends* represent its 13-bp TIRs, whose sequences are shown below the diagram. The *lightly shaded regions* at each end represent the subterminal repetitive sequences which contain TnpA binding sites. The transcription start site at nucleotide 209 is indicated by an *arrow*, and the *darkly shaded block* adjacent to it is the GC-rich DCR. *Arrows within the box* show the locations of the two long continuous ORFs contained within the first intron of the TnpA transcript. The structure of the *Spm* transcripts is shown below the diagram of the element. *Filled blocks* represent the exons and *connecting lines* represent the introns. The structure of the processed transcripts is shown at the *bottom*. Intron locations are represented by *breaks between the filled blocks*, which represent the exons. the *continuous arrows* represent the ORFs encoded in each of the element's transcripts. The transcripts have been designated *tnpA*, *tnpB*, *tnpC* and *tnpD* in order of increasing size. The *tnpA* and *tnpB* transcripts are monocistronic, while the *tnpC* and *tnpD* transcripts are dicistronic, encoding a unique ORF, as well as the ORF encoded by the *tnpA* transcript. The proteins encoded by the single ORFs of the *tnpA* and *tnpB* transcripts have been designated TnpA and TnpB, while those encoded by the ORFs that are the unique to the *tnpC* and *tnpD* transcripts have been designated TnpC and TnpD

frequency and have short deletions within the first intron of the element's most abundant transcript, *tnpA*. The many transposition-defective elements, termed *dSpm* or *I* elements, that have been cloned and analyzed are all derivatives of the *Spm* element with large internal deletions, extending into or eliminating the coding sequence for the element's most abundant mRNA.

The element has 13-bp perfect terminal inverted repeats (TIRs). Adjacent to each TIR is a subterminal internally repetitive region of about 0.2 kb at the 5′ end and 0.3 kb at the 3′ end (Fig. 1). In addition, there is a GC-rich, internally repetitive sequence located between its transcription start site at nucleotide 209 and the end of its untranslated first exon at nucleotide 578. While the TIRs are perfect repeats, the repetitive elements of the subterminal repetitive region and the GC-rich first exon sequence are not all alike. The 13-bp TIR sequence (CACTACAA-GAAAA) differs from the consensus sequence of the subterminal repetitive regions (CCGACACTCTTA) and that of the GC-rich region (CGGGCGGGCGGC-CTCGC). The 3′ and 5′ subterminal repetitive regions share a common repetitive element, of which there are nine copies within the first 0.2 kb at the 5′ and the 14 copies within the 0.3-kb 3′ terminal sequence. The repetitive elements occur in both orientations in the subterminal repetitive regions, while the 11 GC-rich repeats within the first exon are all in the same orientation (MASSON et al. 1987).

2.1 Sequences Required for Transposition

MCCLINTOCK reported that plants with a *dSpm* insertion allele of a gene and a mobilizing intact *Spm* frequently give rise to progeny showing patterns of element excision that differ from the parental pattern (reviewed in FEDOROFF 1983). Mutant progeny differ from the original alleles in the frequency and developmental timing of *dSpm* excision from the gene, as well as the level of gene expression exhibited by the affected allele both in the presence and in the absence of the mobilizing *Spm*. Such changes, which MCCLINTOCK (1955) designated "states" of a locus, are heritable. It is now known that most such changes arise by further deletions within a *dSpm* element and the molecular analysis of such elements has provided information about the sequences required for transposition of the element (reviewed in FEDOROFF 1989a).

Not unexpectedly, all of the sequences that affect the ability of a *dSpm* element to transpose are located at or near element ends. The 13-bp TIRs are important, but will tolerate some changes, as evidenced by the ability of an element with a 2-bp terminal deletion in a *dSpm* element at the *bz* locus to transpose, albeit at reduced frequency (SCHIEFELBEIN et al. 1988). Deletions that extend into the subterminal repetitive sequences also reduce the transposition frequency and delay its timing, as does deletion of the GC-rich repetitive region within the element's first exon (Fig. 1). A single non-transposing element has been identified to date and it contains both TIRs, but lacks the 5′ subterminal region entirely (MENSSEN et al. 1990). The sequence requirements for optimal transposition frequency of the *Spm* element therefore include the element's

TIRs, the subterminal repetitive sequences at both ends, and the GC-rich, repetitive first exon sequence.

3 Expression of the *Spm* Element

A single primary transcript has been identified for the *Spm* element and it extends from nucleotide 209, from what has been designated the element's 5′ end, to one of two transcription termination sites about 0.4 kb from the element's 3′ end (Fig. 1; reviewed in FEDOROFF 1989a). The primary transcript is differentially spliced to give an abundant 2.4-kb transcript, designated *tnpA*, and three minor larger transcripts, designated *tnpB*, *tnpC*, and *tnpD*, in order of increasing size (Fig. 1). The *tnpA* transcript contains 11 exons and ten introns, the first of which is more than 4 kb in length, while the other nine are all less than 0.2 kb in length (PEREIRA et al. 1986). The *tnpB*, *tnpC*, and *tnpD* transcripts each contain all 11 of the *tnpA* transcript's exons, but have one, two, or three additional exons derived from one or both of the open reading frames (ORFs) located within the first intron of the *tnpA* transcript (Fig. 1; MASSON et al. 1989). The first exon, which contains the GC-rich repetitive region, comprises an untranslated leader sequence and the ORFs of all of the transcripts commence in their respective second exons.

The *tnpA* transcript contains a single ORF coding for 68-kDa DNA-binding protein designated TnpA (Fig. 1; PEREIRA et al. 1986; GIERL et al. 1988). The larger *tnpB* transcript likewise has a single ORF which encodes a potential 168-kDa ORF1-ORFA fusion protein designated TnpB. Both *tnpC* and *tnpD* contain two ORFs, one of which is an ORF1-ORF2 fusion, while the second is ORFA. The proteins potentially encoded by the ORF1-ORF2 fusions are unique to the *tnpC* and *tnpD* transcripts and are designated TnpC and TnpD, respectively (Fig. 1). They differ from each other only by the excision of a 96-bp intron from the ORF2 coding sequence of *tnpC* that is part of the *tnpD* third exon (MASSON et al. 1989).

3.1 Element-Encoded Proteins Required for Transposition

The element-encoded proteins that participate in transposition were identified by using cDNAs coding for a single gene product to complement mutant elements and each other in a transgenic assay system (MASSON et al. 1989, 1991; FREY et al. 1990). *Spm* activity was assayed in transgenic tobacco plants by introducing an *Agrobacterium tumefaciens* T-DNA carrying a *dSpm*-disrupted GUS gene together with a second T-DNA carrying an intact *Spm* element (MASSON and FEDOROFF 1989). The activity of the *trans*-activating *Spm* element or cDNAs was judged by the appearance of GUS-positive sectors in transformed calli and plants generated from them. Frameshift mutations in ORFs 1, 2, and A markedly reduced or eliminated the ability of an intact *Spm* element to promote excision of the *dSpm* from the GUS gene. A *tpnA* cDNA was able to complement the defect in elements with frameshift mutations in ORFA, but not ORFs 1 and 2, indicating that both TnpA and either TnpC or TnpD were required for *dSpm* transposition, but that the intact TnpB protein was not required (Masson et al. 1991). None of the cDNAs tested supported excision individually, but the combination of the *tnpA* and *tnpD* cDNAs was found to be both necessary and sufficient to support *dSpm* excision at a high frequency. As described below, TnpA binds to element ends and has been postulated to bring element ends together during transposition. Nothing is presently known about the properties of TnpD, although it has been suggested to be an endonuclease with recognition specificity for *Spm* ends (Masson et al. 1991).

3.1.1 TnpA Is a DNA-Binding Protein

TnpA, the protein encoded by the *tnpA* transcript, binds to the subterminal repetitive regions of *Spm*, but not to its TIRs (Gierl et al. 1988). The target binding sequence has been identified as the 12-bp consensus subterminal repetitive element. TnpA binds as a dimer and the optimal binding site comprises a tail-to-tail dimer of the 12-bp binding site (Trentmann et al. 1993). Binding is reduced if the cytosine residues of the CG dinucleotides and CNG trinucleotides within the binding site are methylated (Gierl et al. 1988). When a dimeric TnpA binding site was inserted between a cauliflower mosaic virus (CaMV) 35S promoter and GUS gene, co-expression of a *tnpA* cDNA in tobacco protoplasts inhibited expression of the GUS gene (Grant et al. 1990). The latter observation suggests that the interaction between the TnpA and the *Spm* termini is sufficiently strong to interfere with transcription from the promoter of a gene into which a *dSpm* element has inserted, providing a possible explanation for McClintock's initial observation that certain *dSpm* insertion mutations have a null phenotype when a nondefective *Spm* element is present in the same genome, by contrast to their ability to express the affected gene at an intermediate level in the absence of a *trans*-acting *Spm* (McClintock 1953, 1954).

3.2 The *Spm* Promoter

The *Spm* promoter was analyzed in transient assays using *Spm* promoter-firefly luciferase gene transcriptional fusions (Cook and Fedoroff 1992; Raina et al. 1993). The constitutive promoter activity resides entirely in the element's 5′ end, termed its upstream control region or UCR (Banks et al. 1988). The GC-rich repetitive first exon, termed the downstream control region or DCR, has no influence on the basal level of *Spm* promoter-driven luciferase expression, but does attenuate the promoter's response to a nearby enhancer sequence (Banks et al. 1988; Cook and Fedoroff 1992; Raina et al. 1993). The UCR contains nine copies of the TnpA binding site (sequences with at least 75% identity to the consensus binding site) and eight of the nine from tail-to-tail dimers (Masson et al. 1987). Consecutively longer deletions extending from the 5′ end of the element through the UCR result in the progressive loss of promoter activity, suggesting that transcription factor binding sites either coincide with or overlap the TnpA binding sites (COOK and FEDOROFF 1992; RAINA et al. 1993).

4 Genetic Studies on *Spm* Inactivation and Reactivation

The *Spm* element can undergo a heritable, but reversible change in its genetic activity (MCCLINTOCK 1957, 1958, 1959, 1961, 1962, 1971). While an inactive *Spm* can neither transpose nor *trans*-activate expression of a *dSpm* element, it is distinguishable from a *dSpm* in two ways: it can return to an active form spontaneously and it can be transiently reactivated by an active element (MCCLINTOCK 1971). Inactive elements differ considerably in the stability of the inactive state, as well as the developmental pattern of element reactivation (reviewed in FEDOROFF 1983, 1989a). In some cases, the element alternates between active and inactive phases frequently during plant development, giving a plant that shows a fine mosaicism for element expression. This type of element has been termed a *cycling Spm* (*Spm-c*). Other elements exhibit heritable developmental patterns of inactivation and reactivation in which the element is active only in certain plant parts or only in certain areas of a given tissue. Still other elements remain largely inactive throughout development, returning to the active phase infrequently and late in development.

At the extreme, an *Spm* element can be so stably inactive that the frequency of spontaneous reactivation approaches the spontaneous mutation frequency (FEDOROFF 1989b). Such stably inactive elements have been termed *cryptic Spm* elements and there is evidence that intact elements persist in the maize genome in this genetically silent form. Thus the *Spm* element can exist in a variety of genetically distinguishable, heritable states. However, since this type of genetic inactivation is reversible, it cannot be heritable in the same sense that MCCLINTOCK's original "states" of a locus were heritable, most of which proved to have sustained internal deletions (MCCLINTOCK 1955). Much of the recent work on the reversible inactivation phenomenon has focused on the mode of inheritance of the inactive state, its molecular correlates, and the conditions under which it can be altered (FEDOROFF and BANKS 1988; FEDOROFF 1989; BANKS and FEDOROFF 1989).

4.1 Genetic Properties of *Spm* Expression Patterns

Spm elements can be divided into broad classes on the basis of the heritability of their activity phase. The *Spm* element can be either active or inactive at the beginning of the plant's developmental cycle and the heritability of the activity phase can be either high or low. Fully active elements rarely undergo spontaneous inactivation early enough in development for the inactivation to be heritable (FEDOROFF and BANKS 1988). The precise frequency is difficult to determine because other genetic changes that affect element activity, such as transposition to a new site and self-promoted internal deletions, occur at a high frequency and often give identical phenotypes. Similarly, the cryptic form is highly heritable. *Cryptic Spm* elements undergo spontaneous reactivation so infrequently that no candidate events were detected in a screen of several hundred thousand kernels carrying a *dSpm* insertion allele capable of detecting the presence of an active element anywhere in the genome (N. Fedoroff, unpublished observations). In addition to the stably active and inactive forms, there are *Spm* elements that regularly undergo one or more activity phase changes, either from active to inactive or vice versa, during plant development, either in a random fashion or in a regular developmental pattern (McClintock 1957; Peterson 1966; Fedoroff and Banks 1988). Such developmental patterns have been termed "programs" of *Spm* expression and elements capable of displaying one or another pattern have been termed "programmable".

4.1.1 *Spm* Expression Programs

Spm elements that undergo changes in activity phase during the developmental cycle have several common properties that distinguish them as a group. The first is that the pattern of phase reversal, defined by its frequency and developmental profile, is heritable. The second is that the heritability of a given pattern is far lower than the heritability of either the stably active or the cryptic state. The third is that the heritability can be altered during development or by the presence of an active element. The number of different developmental patterns of *Spm* expression is probably limited and may be determined by some underlying pattern of expression of certain plant genes, whose number and identity are not known. Indeed, similar developmental patterns of element expression have surfaced in both old and recent studies (McClintock 1957, 1958; Peterson 1966; Fowler and Peterson 1978; Fedoroff and Banks 1988; Fedoroff 1989b). Two of the most common developmental patterns are ones in which the element is more active in ears produced on tillers than in the ears produced on the main stalk of the plant and one in which the element is inactive only in the crowns of kernels. Finally, some patterns are more stable than others, changing heritably at a lower frequency.

4.1.2 An *Spm* Expression Program Is Heritable, but Relatively Unstable

A good illustration of the heritability of an expression pattern and the way in which it changes is provided by an element isolated during studies on the genetic reactivation of *cryptic Spm* elements (Fedoroff 1989b). The line in question contained an element that exhibited behavior similar to that of an element McClintock called *Modifier* because its effect of increasing transposition frequency was discernible only in the presence of an active *Spm* element (McClintock 1957). The newly isolated *Modifier*-like element was also active only in the presence of an *Spm* element, but it was almost never expressed in kernels receiving it through male gametes (Fedoroff 1989b). A few exceptional kernels (<1%), appearing on ears that received the *Modifier*-like element through male gametes, were grown and their progeny analyzed. The *Modifier*-like element in ears produced by the plants showed an altered and somewhat higher transmission frequency through male gametes, as well as *Spm*-like activity in some kernels after segregation from the *Spm* element also present in the culture. While the *Modifier*-like element derived from the exceptional kernels still clearly exhibited the expression pattern of the parental element, its pattern had been heritably altered to more nearly resemble that of a standard *Spm* element (FEDOROFF 1989b). The *Modifier*-like element in plants grown from sibling kernels that showed no alteration had the characteristics of the original element. Thus the properties of the *Modifier*-like element were quite heritable, but proved slightly altered in a small fraction of kernels. The alteration was not all-or-none, but changed the element's expression pattern so that it more nearly resembled that of an *Spm* element. This suggests that both the *Modifier*-like element and its derivatives were stepwise intermediates in the reactivation of a *cryptic Spm*.

4.2 Regulatory Interactions Between *Spm* Elements

MCCLINTOCK (1971) provided the first evidence that an element-encoded gene product is able to alter the element's activity phase. Although the frequency and timing of somatic transposition events is generally not affected by the number of active *Spm* copies in a plant's genome, MCCLINTOCK noted that the pattern of alternation between active and inactive phases changed with element copy number in the case of an *Spm-c* element. MCCLINTOCK used an allele of the *a2* gene in which the inserted *dSpm* element had been immobilized by a subterminal deletion (MENSSEN et al. 1990). This mutant allele of the *a2* gene (*a2-m1*) supports expression of an almost normal level of pigmentation in the absence of *Spm*, but has a null phenotype in the presence of a fully functional element elsewhere in the genome (MCCLINTOCK 1971).

When the *a2-m1* allele is *trans*-activated by an *Spm* element cycling between active and inactive phases (*Spm-c*), pigments are produced only in plant parts in which the element is inactive. When using an *Spm-c* element which undergoes frequent changes in activity phase during development, MCCLINTOCK noted that the non-pigmented area increased with element copy number. She inferred that all copies of the *Spm-c* must be inactive in order for the *a2* gene to be expressed, resulting in pigment synthesis. The differences in the size of the pigmented areas were sufficiently striking to permit determination of the number of resident *Spm* elements. This "dosage effect" proved important in detecting interactions between *Spm* elements. MCCLINTOCK found that when an *Spm* element that had been almost completely inactive in several previous generations was brought together with an *Spm-c* element by a genetic cross, it made a contribution to the dosage effect equivalent to that of an *Spm-c* element. MCCLINTOCK (1971) inferred from this observation that the *Spm-c* element reactivated the inactive element.

4.2.1 An Active *Spm* Transiently Reactivates an Inactive *Spm*

McClintock noted that while two elements having different patterns of inactivation and reactivation were together in a single plant, all of the elements assumed the periodicity of the element showing the most frequent alternation of phases (McClintock 1971). However, the very existence of the dosage effect that permitted McClintock to detect the interaction between elements implies that while all the elements in a plant might assume roughly the same periodicity, the synchronization does not extend to the duration of the activity phase. That is, an active element might reactivate an inactive one, but it does not control its return to the inactive state and the dosage effect arises because more cells contain at least one active element if there are two elements present, each undergoing inactivation independently, than if only one element is present. In subsequent studies, it has become apparent that a fully active element can override a given element's pattern of expression, causing it to be active throughout the developmental cycle of the plant (Fedoroff and Banks 1988; Fedoroff 1989b). However, the response of an inactive *Spm* to an active one depends on the genetic stability of the inactive state. The higher the heritability of the inactive state, the less responsive the inactive *Spm* is to reactivation and *cryptic Spm* elements show no uniform reactivation response.

4.2.2 Active *Spm* Elements Promote the Heritable Reactivation of Inactive Elements

Although an inactive *Spm* element remains in the active phase so long as an active element is present, it generally returns to an inactive form when the *trans*-activating element is removed by meiotic segregation or inactivation (McClintock 1971). Nonetheless, the probability that an inactive element will remain active is increased by exposure to an active element (Banks et al. 1988; Fedoroff 1989b). Thus the introduction and subsequent segregation of a *trans*-activating element substantially increases the fraction of progeny receiving an inactive *Spm* element in a reactivated form.

In addition, the heritability of the inactive phase can change during development. *Spm* elements are commonly transmitted in an active phase more frequently through tiller germlines than through those produced on the main stalk of the plant (McClintock 1957, 1958; Fowler and Peterson 1978; Fedoroff and Banks 1988). However, even when an element is transmitted through a tiller ear in an inactive form, it has a higher probability of being reactivated in the following generation than if it had been transmitted through an ear produced on the main stalk of the same plant (Fedoroff and Banks 1988). This observation implies that genetic changes influencing the heritability of the inactive phase can occur during development and are not necessarily accompanied by a phase change.

4.2.3 Reactivation of *Cryptic Spm* Elements

While the cryptic state is extremely stable and heritable, reactivation of *cryptic Spm* elements has been observed after several types of genetic perturbations. McCLINTOCK discovered transposable elements among progeny of plants which commenced development with two broken chromosomes 9 and which showed evidence of severe abnormalities, both chromosomal and morphological (McCLINTOCK 1945, 1946, 1978). It has also been reported that chromosome-damaging irradiation and passage through tissue culture can activate cryptic elements (NEUFFER 1966; BIANCHI et al. 1969; PESCHKE et al. 1987; PESCHKE and PHILLIPS 1991). Finally, an active *Spm* element can reactivate a cryptic element, albeit slowly.

If an active element is introduced into a strain carrying a *cryptic Spm* element, sectorial activation of the cryptic element can be observed at a low frequency (FEDOROFF 1989b). By carrying out repeated selection of kernels exhibiting sectorial reactivation, fully active elements can eventually be recovered. The reactivation process is stepwise and gradual, taking as long as several plant generations and proceeding through unstable, patterning, readily *trans*-activated intermediates of the type observed in other studies (McCLINTOCK 1957, 1958, 1971; PETERSON 1966; FOWLER and PETERSON 1978; FEDOROFF and BANKS 1988; FEDOROFF 1989b).

Reactivating elements exhibit a phenomenon similar to what McCLINTOCK termed "presetting" in which an element that has been reactivated *in trans* by an active *Spm* exhibits sectorial activity in kernel endosperms immediately following segregation away from the activating element, but is not yet heritably activated (McCLINTOCK 1963, 1964, 1965a, 1967; FEDOROFF 1989b). Plants grown from such kernels produce ears in which occasional sectors of *Spm* activity are detectable by the ability of the element to *trans*-activate excision of a *dSpm* element. Sporadic activity of an inactive element after exposure to an active one is perceptible for one to several generations before the inactive element returns to its previous level of inactivity. These observations underscore the gradual and stepwise genetic changes that lead to the activation of an element in the deeply inactive form designated *cryptic*. The genetic stabilization of the inactive state is similarly a rather slow process, even with selection. Thus, even after four generations of selection for inactivity commencing from an active *Spm* element, it can still be transmitted to a few progeny kernels in an active form (BANKS and FEDOROFF 1989).

5 DNA Methylation and Epigenetic Inactivation of *Spm*

5.1 Methylation Differences and *Spm* Activity in Maize

Changes in methylation of C residues along the length of the *Spm* element were investigated using restriction endonucleases whose ability to cut DNA is inhibited by methylation of C residues in their cleavage sequences (Banks et al. 1988; Banks and Fedoroff 1989). Active and inactive elements alike are methylated throughout most of their sequence, although methylation was not observed to extend into neighboring sequences at the locus studied. Inactive elements could be distinguished from active elements by methylation of sites in the UCR sequence (Banks et al. 1988). *Cryptic Spm* elements could be distinguished from less stably inactive elements by the extent of methylation in the GC-rich DCR. When the methylation of UCR and DCR sites was monitored in a population of plants selected through several generations for inactivity, methylation of the UCR was detectable and increased before that of the DCR (Banks and Fedoroff 1989). These observations imply that methylation of the UCR is correlated with element inactivity, while increasing methylation of sites in the DCR is correlated with the progressive increase in the heritability of the inactive state.

Developmental changes in the genetic activity of an *Spm* element are likewise correlated with changes in element methylation (Banks and Fedoroff 1989). Regardless of the final level of element methylation, reduced levels were observed in embryos. In lines exhibiting more *Spm* reactivation in tillers than in mainstalks, the elements showed lower levels of methylation in the tillers. It has also been observed that the element has a higher probability of being transmitted in an inactive form through male than through female gametes, and this is correlated with higher levels of element methylation in pollen (Fedoroff and Banks 1988; Banks et al. 1988; Banks and Fedoroff 1989). Moreover, the introduction of an active element by a genetic cross decreases the extent of methylation of both inactive and *cryptic Spm* elements (Banks et al. 1988). Consistent with their different responses to transient activation by an active element, inactive elements exhibit lower levels of methylation in the presence of a *trans*-activating *Spm* than do cryptic elements. Thus there is a good correlation between the extent of element methylation and not only its phase of activity, but the heritability of the inactive phase. The more extensively methylated the element's promoter region becomes, the higher the heritability of the inactive phase. The genetic properties of the inactivation and reactivation processes suggest that element inactivity is maintained and rendered more heritable by increases in the number of methylated C residues in the promoter region, particularly in the dense cluster of methylatable C residues in the DCR.

5.1.1 *Spm* Is Inactivated and Methylated in Transgenic Plants

The *Spm* element is active in transgenic tobacco plants (Masson and Fedoroff 1989). *Spm* activity has been monitored in transgenic plants by co-introducing a CaMV 35-*dSpm*-GUS construct. Transgenic tobacco callus tissue and plants are screened for GUS-positive sectors resulting from excision of the *dSpm* elements in the presence of functional *Spm* element or DNAs coding for TnpA and TnpD (Masson and Fedoroff 1989). Transgenic lines containing only a single copy of the *trans*-activating *Spm* element commonly exhibit progressively fewer GUS-positive sectors with time in culture, while lines containing many copies continue to show high levels of *Spm* activity (Schläppi and Fedoroff 1992; Schläppi et al. 1993). The inactivation occurs in callus culture and is stably maintained when plants are regenerated and propagated sexually. Inactivation is accompanied by de novo element methylation. Plants exhibiting few or no somatic GUS-positive sectors show preferential methylation of sites within the element's UCR and DCR sequences.

5.1.2 Methylation Interferes with Transcription and Transposition

Spm elements that are methylated at their 5′ end are not transcribed, although methylation elsewhere within the element does not affect transcription (Banks et al. 1988; Kolosha and Fedoroff 1992; Schläppi et al. 1993). There are also indications that element methylation interferes with the transposition even when TnpA and TnpD are provided *in trans* by an active *Spm*. A full length, but extensively methylated, *Spm* element transposes as poorly as an element that is missing the DCR or part of one or both subterminal repetitive regions, even in the presence of a fully active *Spm* (Fedoroff et al. 1988; Banks et al. 1988). Moreover, an internally deleted element that retains the DCR and normally transposes frequently and early in development in the presence of a *trans*-activating *Spm* transposes much less frequently and later in development when methylated (Schiefelbein et al. 1988; V. Raboy, personal communication). This is not surprising in view of the fact that TnpA is required for element transposition (Frey et al. 1990; Masson et al. 1991) and shows a lower affinity to its binding sequence when the internal C residues are methylated (Gierl et al. 1988).

5.2 The *Spm* Promoter Is the Target of De Novo Methylation

The preferential methylation of the *Spm* element's UCR and DCR sequences in tobacco suggested these as targets of methylation. To determine whether the *Spm*'s promoter sequence was sufficient for de novo methylation in transgenic plants, *Spm*-promoter-firefly luciferase transcriptional fusions were introduced into tobacco by *Agrobacterium*-mediated transformation (Schläppi et al. 1994). The levels of luciferase gene expression were higher by an order of magnitude when the luciferase gene was expressed from a UCR promoter only than when the DCR sequence was included in the promoter. This is in marked contrast to the observation that UCR and UCR + DCR promoters have equivalent capacities to support luciferase gene expression in a transient assay (RAINA et al. 1993). Substantially more extensive methylation of the UCR was associated with lower ability of the UCR + DCR promoter than the UCR-only promoter to support luciferase gene expression in transgenic plants (SCHLÄPPI et al. 1994). These experiments established that the *Spm*'s promoter region is a sufficient target for rapid de novo methylation in transgenic plants. Moreover, they revealed that the presence of the GC-rich DCR sequence in some way accelerates de novo methylation of the *Spm* promoter.

6 TnpA Is a Novel Epigenetic Regulatory Protein

6.1 TnpA Represses an Unmethylated *Spm* Promoter

The effect of TnpA on expression of the unmethylated *Spm* promoter has been studied in transient assays of sufficiently short duration to preclude extensive DNA replication-associated methylation. A cDNA coding for TnpA was expressed from promoters of different strengths and introduced into cultured tobacco cells with an *Spm*-promoter-luciferase transcriptional fusion by particle bombardment (COOK and FEDOROFF 1992; SCHLÄPPI et al. 1994). In this type of assay, TnpA was found to repress expression of the *Spm* promoter comprising either the UCR alone or both the UCR and DCR sequences. Moreover, co-expression of TnpA with the *Spm*-luciferase fusion invariably decreased the level of luciferase activity, although repression decreased with decreasing strength of the promoter driving TnpA.

Similar experiments were carried out in stably transformed cells containing a hypomethylated UCR-luciferase transcriptional fusion (SCHLÄPPI et al. 1994). A *tnpA* cDNA was introduced into such plants by retransformation. Internal control constructs were included to permit the independent monitoring of the intracellular TnpA supply by its ability to promote excision of a *dSpm* element from a 35S-GUS gene. The results of these experiments revealed that TnpA repressed expression of luciferase from the hypomethylated promoter at high intracellular levels of TnpA. Taken together, these observations suggest that TnpA represses the *Spm* promoter if it is unmethylated.

6.2 TnpA Activates the Methylated *Spm* Promoter

6.2.1 TnpA Reactivates an Inactive *Spm* Element

As described earlier, active *Spm* elements promote the reactivation of inactive elements in maize. The element-encoded protein responsible for element reactivation was identified by introducing each of the element's four potential coding sequences into transgenic tobacco plants with an inactive, methylated *Spm* element (SCHLÄPPI et al. 1993). Activity of the inactive *Spm*, as judged by the appearance of many large GUS-positive sectors, is restored by the introduction of *tnpA* cDNA either by retransformation or by a genetic cross. None of the other element-encoded cDNAs had any effect on the activity of the resident *Spm* element. Genetic reactivation, as judged by the GUS assay, is correlated with increase element transcription and decreased methylation of the UCR and DCR sequences. *Spm* activity declines and promoter methylation levels generally increase upon segregation of the *tnpA* cDNA from the element. However, several instances resembling "presetting" have also been observed in which the element remained active and hypomethylated in the generation following segregation of the activating *tnpA* cDNA (SCHLÄPPI et al. 1993).

6.2.2 TnpA Activates an Inactive *Spm* Promoter

The effect of TnpA on the methylated UCR + DCR promoter was examined in stably transformed tobacco cells containing a UCR + DCR-luciferase construct (SCHLÄPPI et al. 1994). As noted earlier, the *Spm* promoter is rapidly methylated in such cells and the luciferase gene is expressed at very low levels. A *tnpA* cDNA was introduced into such cells either by retransformation or by a genetic cross. Internal control constructs were included to monitor the TnpA concentration by its ability to complement TnpD in promoting excision of a *dSpm* from a 35S-GUS gene. Expression of the luciferase gene in such experiments was proportional to the supply of TnpA, as judged by the excision assay. Moreover, promoter methylation decreased with increasing luciferase expression. Thus TnpA promotes both activation and demethylation of an inactive, methylated *Spm* promoter.

Whether TnpA activates the methylated *Spm* promoter directly or indirectly by interfering with methylation is not yet known. Preliminary observations suggest that low levels of promoter activation are observed when a *tnpA* cDNA is introduced by particle bombardment into cells containing an inactive, methylated *Spm* promoter. However, these experiments are difficult to interpret because the effect is small, either because only a small fraction of the cells receive and express the introduced cDNA or perhaps because TnpA is eliciting a response from some cells and not others. Thus the question of the mechanism by which TnpA activates the methylated *Spm* promoter remains open.

6.3 Deletion Studies on TnpA

Oligonucleotide-binding studies were used to identify the regions of the TnpA protein required for DNA binding and for dimerization (TRENTMANN et al. 1993). In these studies, a DNA-binding domain was identified between amino acids 122 and 427 of the 621 amino acid TnpA protein, while a protein-dimerization domain was localized between amino acids 428 and 542. Studies were carried out to identify the regions of the protein required for its transcriptional repression and activation functions (SCHLÄPPI et al. 1994). The results of these studies showed that only the DNA-binding and dimerization domains are required for transcriptional repression of the unmethylated *Spm* promoter by TnpA. A deletion derivative of TnpA lacking the N-terminal 122 amino acids was an equivalent or better transcriptional *Activator* of the methylated *Spm* promoter, while deletion derivatives lacking even the C-terminal 81 amino acids were no longer active. This suggests that the TnpA protein's repressor and activation functions are either mechanistically different or require the participation of additional protein co-factors.

7 Summary and Conclusions

7.1 TnpA Is a Multifunctional Epigenetic Regulatory Protein

7.1.1 TnpA Binding Represses Transcription and Promotes Transposition

TnpA, one of the two proteins potentially encoded by the *Spm* element for which functions have been identified, is both a transcriptional regulatory protein and a protein that participates directly in transposition. The multiple functions of TnpA are reflected in the sequence organization of the element. The element's promoter sequence is short and coincides with the 5′ subterminal repetitive sequence, more than half of which is occupied by nine TnpA binding sites. Promoter activity is progressively lost as TnpA binding sites are deleted, suggesting that there are multiple binding sites for transcription factors coincident with or overlapping the TnpA binding sites. The observation that only the DNA binding and dimerization domains of TnpA are required for promoter repression suggests that TnpA's repressor activity is a direct consequence of its ability to bind to the element's promoter and displace transcription factors.

Although little is known about the role of TnpA in transposition, its ability to bind more tightly to the element's promoter than transcription factors may be important in the transition from a transcriptionally active element to a transcriptionally competent element. It has been suggested that TnpA serves to bring element ends together prior to transposition, but direct evidence for this is lacking (Frey et al. 1990; Masson et al. 1991). However, deletion of even a few of its binding sites at either element end delays transposition and decreases its frequency (Schiefelbein et al. 1985; Schwarz-Sommer et al. 1985). This suggests that most or all of the TnpA binding sites must be present and occupied for optimal transposition of the element. Because the binding sites are not all identical in sequence, it may be that transposition occurs only at concentrations of TnpA sufficient to assure saturation of binding sites with both low and high TnpA affinities (Fig. 2). Supporting this interpretation is the observation that inhibition of luciferase expression from a hypomethylated promoter in transgenic plants was observed only in plants with high levels of TnpA, as judged by its ability to stimulate transposition (Schläppi et al. 1994).

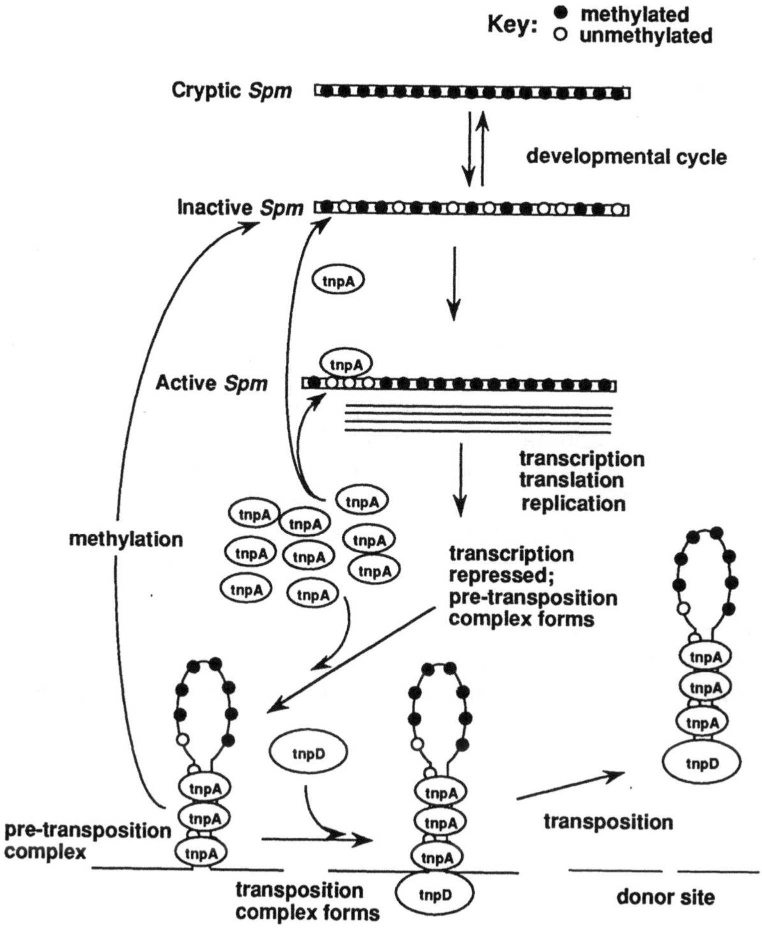

Fig. 2. A model of *Spm* regulation. The *open boxes* represent *Spm* elements. *Open* and *filled circles* represent unmethylated and methylated sites, respectively. *Spm* elements are represent as undergoing changes in the extent of methylation during the developmental cycle, as described in the text, which results in reduced methylation early in development, followed by increasing methylation during development in a tissue-specific manner. TnpA is represented as interacting with the element's 5′ end to maintain it in a transcriptionally active hypomethylated state relative to the remainder of the element. Accumulation of TnpA is postulated to inhibit transcription of the element and promote formation of a pre-transposition complex in which the element ends are associated with each other through TnpA-mediated interactions. Conversion of the pre-transposition complex to the transpositionally competent form is postulated to occur with the addition of TnpD to the complex, followed by transposition to a new site. An occasional alternative fate of the pre-transposition complex is proposed to be dissociation, releasing the element in a methylation-sensitive form

7.1.2 TnpA Is a Novel Epigenetic Activator Protein

The ability of TnpA to activate and promote demethylation of the *Spm* promoter identifies it as the first member of a new class of epigenetic regulatory proteins. Whether TnpA activates the methylated promoter directly or indirectly by reducing methylation remains poorly understood. Clearly both may play a role in the ability of TnpA to effect transcriptional reactivation. It is known that when a heavily methylated *cryptic Spm* element is exposed to an active TnpA-producing *Spm*, a substantial reduction in methylation can be observed without concomitant evidence of either increased element transcription or element transposition (BANKS et al. 1988). Conversely, some promoter methylation is compatible with continued promoter expression, and, as noted above, promoter repression occurs only at high TnpA levels in transgenic plants. In view of the fact that TnpA binds less strongly to sites containing methyl groups, the interaction between TnpA and the *Spm* promoter may be governed by *both* the extent of promoter methylation *and* the concentration of the protein.

One of the most interesting aspects of TnpA's interaction with methylated *Spm* promoter is its ability to promote demethylation of sequences outside of its binding sites, suggesting that the mechanism by which it influences promoter methylation is not a simple competition with methylase for binding to its recognition site. Only 12 of the more than 100 methylatable C residues on each DNA strand in the UCR and DCR are within TnpA binding sites. The interaction between TnpA and the methylated promoter results in decreased methylation of the entire regions. This may be attributable to the ability of the TnpA to maintain the DNA in a configuration that is less favorable or accessible to maintenance of methylation during DNA replication. Or its effect may be mediated by components of the transcription machinery. That is, TnpA binding may promote binding of other transcription factors, which in turn decreases the probability of remethylation. Whatever its mechanism of action, it is clear that TnpA can have only a modest influence on the extent of element remethylation at any given round of DNA replication. This follows from the observation that even in the presence of a source of TnpA and even under selection, multiple plant generations are required for the conversion of a *cryptic Spm* element to a fully active one (FEDOROFF 1989b).

7.2 Methylation and *Spm* Element Expression

The relationship between *Spm* element expression and methylation is complex. Although it is often asked whether promoter methylation is causal or consequential to element inactivation, it may be both. The *Spm* promoter is extremely weak and contains many methylatable C residues, making it a likely target for preferential de novo methylation (BANKS et al. 1988; RAINA et al. 1993; SCHLÄPPI et al. 1993). Thus methylation may occur because the promoter binds transcription factors very weakly and methylation may, in turn, weaken the binding further, facilitating more extensive methylation to lock in the inactive phase. At the same time, there are developmental changes in element methylation and the levels of element methylation are low early in development. This may facilitate the transcriptional activation of an element that is not extensively methylated and account for differential expression of the element during development (BANKS and FEDOROFF 1989). However, the observation that the extensively methylated cryptic from of the element does not become completely demethylated in the promoter region either early in development or in the presence of a *trans*-acting *Spm* suggests that while methylation is reduced under these circumstances, it is not entirely eliminated (BANKS et al. 1988; BANKS and FEDOROFF 1989).

Although a *cryptic Spm* cannot be fully reactivated by a *trans*-acting *Spm*, rare kernels nonetheless exhibit the reactivated phenotype and the cryptic element in such kernels has been heritably altered (FEDOROFF 1989b). This observation suggests that the elements in a plant are not a homogeneous, but rather a heterogeneous population with respect to the extent of promoter methylation. Support for this inference is provided by the observation that different plants containing both a *cryptic Spm* and a *trans*-activating *Spm* exhibit considerable variation in the extent to which a given site within the promoter is methylated (BANKS et al. 1988). The *Spm*'s promoter region contains many methylatable sites and all sites tested so far have been found to be almost completely methylated in plants with *cryptic Spm* elements. By contrast, in plants with less stably inactive elements, a given element may or may not be methylated at a particular site in the DCR (BANKS et al. 1988). Finally, it takes several generations of selection to derive a stably active element from an inactive one or a stably inactive element from an active one (BANKS and FEDOROFF 1989; FEDOROFF 1989b). Developmental patterns of element expression are observed with elements whose activity phase is not highly heritable, and the patterns themselves are heritable but relatively labile (FEDOROFF and BANKS 1988; FEDOROFF 1989b). Moreover, methylation differences mirror the element's developmental pattern of expression (BANKS and FEDOROFF 1989). Thus a given developmental pattern of *Spm* expression appears to be the result of the interactions between the germinally transmitted extent or pattern of *Spm* promoter methylation and the methylation changes that occur during development. The observation that developmental patterns are heritable suggests that the level or pattern of element methylation in one generation is reproduced in the next generation. At the same time, the population of elements transmitted through the germline exhibits considerable heterogeneity in both methylation and genetic activity, providing the potential for selecting either more or less active lines. the efficacy of selection implies that the heterogeneity is genetic. The relatively slow response time for selecting a fully inactive line is consistent with the apparent multisite requirement and the heterogeneity introduced in each generation.

7.2.1 Methylation and the Evolutionary Survival of *Spm*

The ability of the *Spm* element to exist in a cryptic form may be essential for its long-term survival in the genome. Active transposable elements cause a variety of chromosomal rearrangements and these are often lethal. Moreover, active elements may contribute to their own demise, promoting intraelement deletions (MASSON et al. 1987). The stability of the cryptic state may find its explanation in the DCR sequence with its multiplicity of methylatable C residues. TnpA binding sites are confined to the UCR, yet the heritability of the inactive states is correlated with the extent of DCR methylation. When an element is inactivated, methylation of the UCR is detectable before methylation in the DCR, suggesting that methylation of TnpA binding sites suffices to inactivate the promoter (BANKS and FEDOROFF 1989). But the inactive state becomes highly heritable only when the DCR is fully methylated. While the role of the DCR in maintaining the heritability of the inactive state is not well understood, it may be that it simply has enough methylated sites so that developmental demethylation extensive enough to permit spontaneous transcriptional reactivation of the element is extremely unlikely.

Although the reason is not yet known, it is clear that properties of *Spm*'s promoter sequence make it a good target for de novo methylation, suggesting that the propensity to become methylated might reside in the nucleotide sequence of the element itself. At the same time, *Spm* encodes a protein one of whose functions is to interfere with methylation and transcriptional inactivation of the element. The question of causality, therefore, appears less fruitful than the question of why transposable elements have such a mechanism of heritable, reversible inactivation. The answer may lie in the paradox that transposition itself endangers the survival of a transposable element, simply because of the ever-present potential of causing genetic damage during transposition or by the disruption of an essential gene. Transposable elements in other organisms possess a hierarchy of mechanisms that redundantly ensure the maintenance of a low transposition frequency (reviewed in BERG and HOWE 1989) and it appears that at least some of the maize elements do, as well. All of the maize transposable elements studied so far exhibit the property of heritable, reversible genetic inactivation associated with methylation of certain element sequences.

As is evident from other chapters in the present volume, plants commonly methylate sequences present in multiple copies as a consequence of either duplication within the genome or exogenous introduction as transgenes. The potential for increasing the redundancy of transposable element sequences in the genome is inherent in the transposition process. Thus, regardless of an element's primary sequence, inactivation by methylation may be an unavoidable occupational hazard for transposable elements residing in plant genomes. From this perspective, it is not surprising to find that plant transposable elements have evolved mechanisms that interfere with the inactivating effects of methylation. At the same time, it is increasingly evident that all of the maize transposable elements have a tendency to undergo self-promoted intraelement deletions. Thus elements have not only the potential of creating lethal mutations, but also a marked tendency towards self-destruction. There is substantial genetic and molecular evidence that plant transposable elements exist in most genomes in a cryptic, heavily methylated state. Thus it may be that the long-term genetic survival of transposable elements is actively favored by the existence and properties of plants' general methylation mechanisms.

References

Banks JA, Fedoroff N (1989) Patterns of developmental and heritable change in methylation of the Suppressor-mutator transposable element. Dev Genet 10: 425–437

Banks JA, Masson P, Fedoroff N (1988) Molecular mechanisms in the developmental regulation of the maize Suppressor-mutator transposable element. Genes Dev 2: 1364–1380

Bennetzen JL (1987) Covalent DNA modification and the regulation of Mutator element transposition in maize. Mol Gen Genet. 208: 45–51

Berg D, Howe M (1989) Mobile DNA. American Society for Microbiology, Washington DC

Bianchi A, Salamini F, Restaino F (1969) Concomitant occurrence of different controlling elements. Maize Genet Coop News Lett 43: 91

Chandler V, Walbot V (1986) DNA modification of a maize transposable element correlates with loss of activity. Proc Natl Acad Sci USA 83: 1767–1771

Cook D, Fedoroff N (1992) Regulation of Spm promoter activity by the Spm-encoded TnpA gene product and DNA methylation. Maize Genet Coop Newslet 66: 11–12

Cone CC, Schmidt RJ, Burr B and Burr FA (1988) Advantages and limitations of using Spm as a transposon tag. In: Nelson O (ed) Plant transposable elements. Plenum, New York pp 149–159

Fedoroff NV (1983) Controlling elements in maize. In: Shapiro J (ed) Mobile genetic elements. Academic, New York, pp 1–63

Fedoroff NV (1989a) Maize transposable elements. In: Berg D, Howe M (eds) Mobile DNA. American Society for Microbiology, Washington DC, pp 375–411

Fedoroff NV (1989b) The heritable activation of cryptic Suppressor-mutator elements by an active element. Genetics 121: 591–608

Fedoroff NV, Banks JA (1988) Is the Suppressor-mutator element controlled by a basic development regulatory mechanism? Genetics 120: 559–577

Fedoroff NV, Wessler S, Shure M (1983) Isolation of the transposable maize controlling elements Ac and Ds. Cell 35: 243–251

Fedoroff NV, Masson P, Banks J, Kingsbury J (1988) Positive and negative regulation of the Suppressor-mutator element. In: Nelson O (ed) Plant transposable elements. Plenum, New York, pp 1–15

Fowler RB, Peterson PA (1978) An altered state of a specific En regulatory element induced in a maize tiller. Genetics 90: 761–782

Frey M, Reinecke J, Grant S, Saedler H, Gierl A (1990) Excision of the En/Spm transposable element of Zea mays requires two element-encoded proteins. EMBO J 9: 4037–4044

Gierl A, Lütticke S, Saedler H (1988) TnpA product encoded by the transposable element En-1 of Zea mays is a DNA binding protein. EMBO J 7: 4045–4053

Grant SR. Gierl A, Saedler H (1990) En/Spm encoded tnpA protein requires a specific target sequence for suppression. EMBO J 9: 2029–2035

Kolosha V, Fedoroff N (1992) Transcriptional reactivation of inactive Spm elements in the presence of Spm-w elements. Maize Genet Coop Newslett 66: 9–11

Masson P, Fedoroff N (1989) Mobility of the maize Suppressor-mutator element in transgenic tobacco cells. Proc Acad Sci USA 86: 2219–2223

Masson P, Surosky R, Kingsbury J, Fedoroff NV (1987) Genetic and molecular analysis of the Spm-dependent a-m2 alleles of the maize a locus. Genetics 117: 117–137

Masson P, Rutherford G, Banks JA, Fedoroff NV (1989) Essential large transcripts of the maize Spm transposable element are generated by alternative splicing. Cell 58: 755–765

Masson P, Strem M, Fedoroff N (1991) The tnpA and tnpD gene products of the Spm element are required for transposition in tobacco. Plant Cell 3: 73–85

McClintock B (1945) Cytogenetic studies of maize and Neurospora. Carnegie Inst Wash Yrbk 44: 108–112

McClintock B (1946) Maize genetics. Carnegie Inst Wash Yrbk 45: 176–186

McClintock B (1953) Mutation in maize. Carnegie Inst Wash Yrbk 52: 227–237

McClintock B (1954) Mutations in maize and chromosomal abberrations in Neurospora. Carnegie Inst Wash Yrbk 53: 254–260

McClintock B (1955) Controlled mutation in maize. Carnegie Inst Wash Yrbk 54: 245–255

McClintock B (1957) Genetic and cytological studies of maize. Carnegie Inst Wash Yrbk 56: 393–401

McClintock B (1958) The suppressor-mutator system of control of gene action in maize. Carnegie Inst Wash Yrbk 57: 415–452

McClintock B (1959) Genetic and cytological studies of maize. Carnegie Inst Wash Yrbk 58: 415–429

McClintock B (1961) Further studies of the suppressor-mutator system of control of gene action in maize. Carnegie Inst Wash Yrbk 60: 469–476

McClintock B (1962) Topographical relations between elements of control systems in maize. Carnegie Inst Wash Yrbk 61: 448–461

McClintock B (1963) Further studies of gene-control systems in maize. Carnegie Inst Wash Yrbk 62: 486–493

McClintock B (1964) Aspects of gene regulation in maize. Carnegie Inst Wash Yrbk 63: 592–602

McClintock B (1965a) Components of action of the regulators Spm and Ac. Carnegie Inst Wash Yrbk 64: 527–534

McClintock B (1965b) The control of gene action in maize. Brookhaven Symp Biol 18: 162–184

McClintock B (1967) Regulation of pattern of gene expression by controlling elements in maize. Carnegie Inst Wash Yrbk 65: 568–578

McClintock B (1971) The contribution of one component of a control system to versatility of gene expression. Carnegie Inst Wash Yrbk 70: 5–17

McClintock B (1978) Mechanisms that rapidly reorganize the genome. Stadler Genet Symp 10: 25–48

Menssen A, Höhmann S, Martin W, Schnable PS, Peterson PA, Saedler H, Gierl A (1990). The En/Spm transposable element of Zea mays contains splice sites at the termini generating a novel intron from a dSpm element in the A2 gene. EMBO J 9: 3051–3057

Neuffer MG (1966) Stability of the suppressor element in two mutator systems at the A1 locus in maize. Genetics 53: 541–549

Pereira A, Cuypers H, Gierl A, Schwarz-Sommer Z, Saedler H (1986) Molecular analysis of the En/Spm transposable element system of Zea mays. EMBO J 5: 835–841

Peschke VM, Phillips RL (1991) Activation of the maize transposable element Suppressor-mutator (Spm) in tissue culture. Theor Appl Genet 81: 90–97

Peschke VM, Phillips RL, Gengenbach BG (1987) Discovery of transposable element activity among progeny of tissue culture-derived maize plants. Science 238: 804–807

Peterson PA (1966) Phase variation of regulatory elements in maize. Genetics 54: 249–266

Raina R, Cook D, Fedoroff N (1993) The maize Spm transposable element has an enhancer-insensitive promoter. Proc Natl Acad Sci USA 90: 6355–6359

Schiefelbein JW, Raboy V, Fedoroff NV, Nelson OE (1985) Deletions within a defective Suppressor-mutator element in maize affect the frequency and developmental timing of its excision from the bronze locus. Proc Natl Acad Sci USA 82: 4783–4787

Schiefelbein JW, Raboy V, Kim H-Y, Nelson OE (1988) Molecular characterization of Suppressor-mutator (Spm)-induced mutations at the bornze-1 locus. In: Nelson O (ed) Plant transposable elements. Plenum, New York, pp 261–278

Schläppi M, Fedoroff N (1992) Promotion of early Spm transposition and repression of Spm transcription by TnpA in transgenic tobacco. Maize Genet Coop Newslett 66: 12–14

Schläppi M, Smith D, Fedoroff N (1993) TnpA trans-activates methylated maize Suppressor-mutator transposable elements in transgenic tobacco. Genetics 133: 1009–1021

Schläppi M, Raina R, Fedoroff N (1994) Epigenetic regulation of the maize Spm transposable element: novel activation of a methylated promoter by TnpA. Cell 77: 427–437

Schwarz-Sommer Z, Gierl A, Berndtgen R, Saedler H (1985) Sequence comparison of 'states' of a1-m1 suggests a model of Spm (En) action. EMBO J 4: 2439–2443

Trentmann SM, Saedler H, Gierl A (1993) The transposable element En/Spm encoded TNPA protein contains a DNA binding and dimerization domain. Mol Gen Genet

Genetic and Epigenetic Inactivation of Repetitive Sequences in *Neurospora crassa*: RIP, DNA Methylation, and Quelling

M.J. Singer and E.U. Selker

1 Introduction

The genome of the filamentous fungus *Neurospora crassa* is approximately 100-fold smaller than that of vertebrates, and contains little repetitive DNA. The existence of repeat-induced point mutation (RIP; reviewed in Selker 1990b) and quelling (Pandit and Russo 1992; Romano and Macino 1992) may account for the rarity of repetitive sequences. RIP is a dramatic example of gene inactivation: duplicated gene-sized DNA sequences are efficiently detected and peppered with transition mutations in the sexual phase of the *Neurospora* life cycle. Asexual inactivation of repeated genes, termed quelling, has also been observed in some *Neurospora* transformants. DNA methylation is often associated with quelling and RIP, but methylation has not been demonstrated to be a requirement for these processes. Many eukaryotic organisms can silence repetitive genes. The phenomena in higher plants referred to as *trans*-inactivation, co-suppression,

Institute of Molecular Biology, University of Oregon, Eugene, OR 97403, USA

and RIGS (repeat induced gene silencing), reviewed in MATZKE and MATZKE (1993) and in other chapters of this volume, show similarities to RIP and quelling. It is still unclear, however, whether the phenomena in plants are mechanistically related to gene inactivation processes operating in fungi.

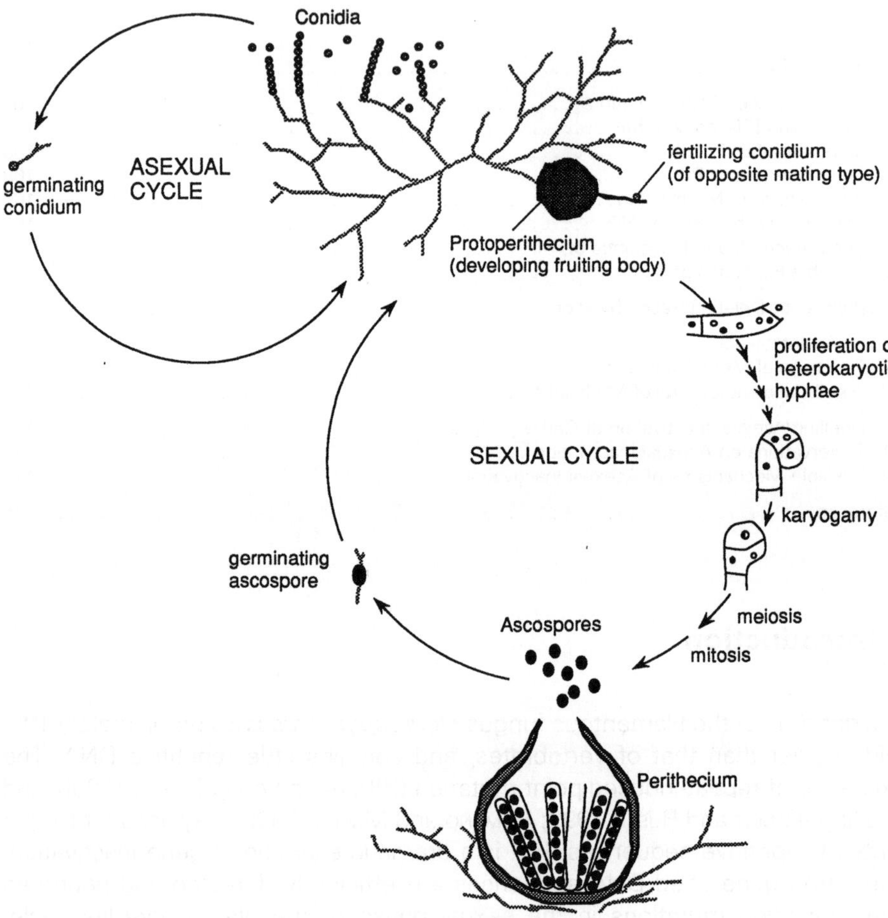

Fig. 1. Life cycle of *Neurospora crassa*. Germination of either a vegetative spore (conidium) or a sexual spore (ascospore) initiates the asexual cycle. Growth of the germling produces branched, multinucleate hyphal cells, and maturation results in the characteristic orange conidia. Nitrogen starvation induces production of fruiting body precursors (protoperithecia). A haploid nucleus of the opposite mating type enters the protoperithecium via a specialized structure and fertilizes a single cell. The resultant heterokaryon proliferates in the enlarging perithecium. Dikaryotic cells containing one nucleus of each mating type (represented by *filled* and *unfilled circles*) at the crest of hook-shaped croziers undergo premeiotic DNA synthesis and nuclear fusion (karyogamy) to give the only diploid cell in the life cycle (diploid nucleus represented by *half-filled circle*). Meiosis and one mitotic division immediately follow, leading to the development of ordered eight-spored asci. (Reproduced with permission from SELKER 1990b, copyright 1990, by Annual Reviews Inc.)

2 Repeat-Induced Point Mutation

2.1 Timing and Efficiency of RIP

Repetitive DNA is highly unstable in the sexual cycle of *Neurospora* (SELKER et al. 1987), but duplications are not affected by RIP in vegetative cells. RIP occurs in the haploid nuclei of special dikaryotic cells that proliferate during the period between fertilization and karyogamy (Fig. 1). Multiple rounds of RIP are thought to occur during seven or more replication cycles that precede premeiotic DNA synthesis and karyogamy (PERKINS and BARRY 1977; SELKER et al. 1987). Diploidy is limited to the zygote, which undergoes meiosis immediately after it is formed.

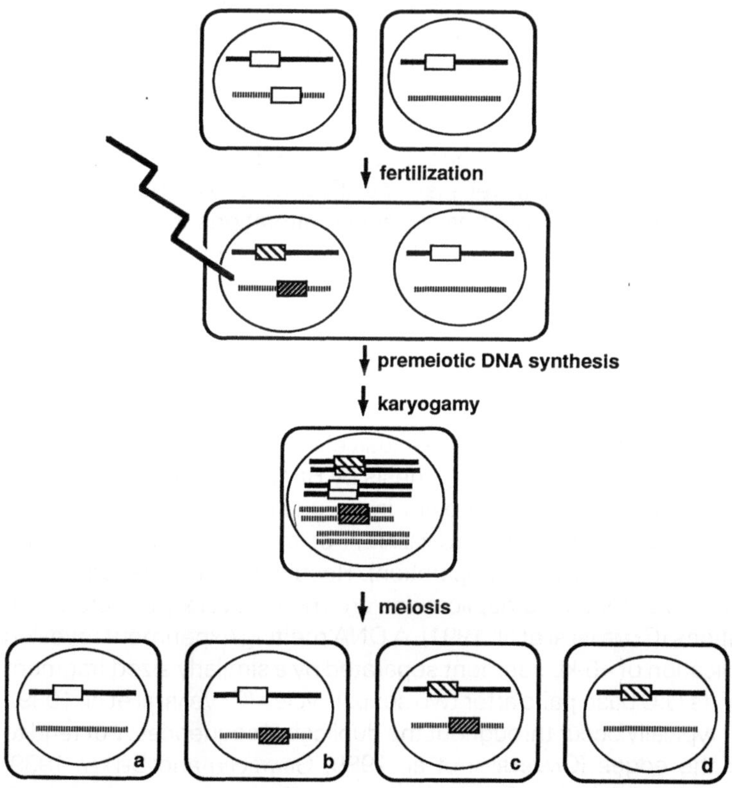

Fig. 2. RIP in an unlinked duplication. Two haploid strains of opposite mating type are illustrated, one containing an unlinked duplication. For simplicity, only two chromosomes are shown. *Open boxes* represent a generic wild-type DNA sequence. *Filled boxes* represent mutated DNA sequences. The *lightning bolt* emphasizes the time in which RIP occurs. The four possible combinations of chromosomes in the progeny are represented as *a-d*. If the duplicated segment contains an essential gene, *c* and *d* are expected to be inviable. DNA methylation can occur in the RIP-altered sequences in *b, c,* and *d*

Because RIP is nucleus limited, a strain containing a duplicated essential gene can produce some viable offspring (Fig. 2).

Repeat-induced point mutation is highly efficient. Tandem duplications of gene-sized sequences rarely escape RIP (SELKER et al. 1987; BOWRING and CATCHESIDE 1993). Unlinked gene duplications are altered at frequencies of 10%–70% (SELKER and GARRETT 1988; Singer and Selker, unpublished; IRELAN et al. 1994). The studies cited above involved genes greater than 1.5 kb in length, but RIP has been reported in linked duplicated sequences as short as 383 bp (STADLER et al. 1991) and unlinked duplicated sequences as short as ≈1kb (SELKER et al. 1989).

2.2 Substrate of RIP

DNA duplications per se activate RIP (SELKER and GARRETT 1988). Single-copy genes are not affected by RIP, and RIP is not merely a response to integration of new DNA (or methylated DNA) into *Neurospora* chromosomes. Furthermore, RIP appears to affect either both or neither elements of a duplication (SELKER and GARRETT 1988), which suggests that duplicated sequences are altered in a pairwise fashion. Support for this conclusion came from investigations of RIP in nuclei containing three copies of a gene. In these experiments, zero, two, or three copies were affected, but never just one (FINCHAM et al. 1989). Multiple rounds of pairwise interactions could have caused alterations in three copies. These results suggest that RIP involves a DNA-DNA interaction between homologous sequences.

2.3 Consequences of RIP

Repeat-induced point mutation results exclusively in G:C to A:T mutations (CAMBARERI et al. 1989). The level of mutagenesis by RIP is variable; less than 1% to greater than 10% of the G:C base pairs in an unlinked duplication can be mutated during a single sexual cycle (CAMBARERI et al. 1989, 1991; JARAI and MARZLUF 1991; SINGER and SELKER, in preparation). Results of hybridization experiments suggest that closely linked duplications are more severely affected than unlinked duplications (CAMBARERI et al. 1991). A DNA melting experiment revealed that a linked duplication of ≈6-kb segment separated by a similarly sized fragment had lost ≈50% of its G:C base pairs after two sexual cycles (CAMBARERI et al. 1989). Mutations of RIP typically occur throughout the duplicated sequences, but tend to be sparser near the edges (CAMBARERI et al. 1989; GRAYBURN and SELKER 1989; Singer and Selker, unpublished). Mutations occasionally occur adjacent to duplicated sequences, however (FOSS et al. 1991; IRELAN et al. 1994). The general coextension of mutations and duplicated sequences is consistent with the idea that RIP involves a DNA-DNA interaction. Analysis of the local sequence context of mutations revealed that RIP has a sequence preference at the dinucleotide level: approximately 74% of affected cytosines are 5′ to adenines, while approximately 14%, 11%, and 1% are 5' to thymines, guanines, and other cytosines, respectively. The impact of RIP on a gene can be severe, and indeed, *Neurospora* researchers have used RIP to disrupt many genes corresponding to cloned sequences (e.g. BELL et al. 1992; CONNERTON 1990; FINCHAM 1990; GLASS and LEE 1992; HARKNESS et al. 1994; JARAI and MARZLUF 1991; MARATHE et al. 1990; NELSON and METZENBERG 1992; SELKER et al. 1989; SELKER and GARRETT 1988).

2.4 RIP-Associated Methylation

Overall, about 1.5% of the cytosines are methylated in *N. crassa* DNA (FOSS et al. 1993; RUSSELL et al. 1985). The bulk of the genome appears completely free of methylation, but several heavily methylated chromosomal loci are known (RUSSELL et al. 1985; SELKER and STEVENS 1985). Most, or possibly all, of the methylated regions appear to be relics to RIP (GRAYBURN and SELKER 1989; Selker, Tauntas, Margolin, Cross and Bird, unpublished). Heavy methylation occurs in the majority of DNA segments altered by RIP (SELKER et al. 1987, 1993a; SELKER and GARRETT 1988; SINGER and SELKER, in preparation). Nearly 100% of the cytosines in a duplicated segment can be methylated (SELKER et al. 1993a). The degree and distribution of the methylation is generally correlated with the degree and distribution of the mutations (SINGER and SELKER, in preparation). However, methylated cytosines can be found beyond the duplication boundary and hundreds of base pairs from the nearest mutation (SELKER et al. 1993a). No correlation between methylation and the number of gene copies is obvious in vegetative cells. Although a duplication is required to trigger RIP, any resulting DNA methylation can persist even after elements of the duplication segregate in meiosis (SELKER and GARRETT 1988). DNA-mediated transformation experiments demonstrated that sequences altered by RIP typically direct de novo methylation (SELKER et al. 1987b; SINGER and SELKER, in preparation). Cloned DNA sequences, demethylated by amplification in *Escherichia coli*, became faithfully remethylated after introduction into vegetative *Neurospora* cells. In contrast, unmethylated *Neurospora* sequences, such as wild-type genes, generally did not become methylated. Thus, some or all of the RIP-associated methylation is a response to the new sequences created by RIP.

2.5 Possible Mechanisms of RIP

The pairwise action of RIP and the coextension of mutations and duplicated sequences are evidence that RIP involves DNA pairing. This pairing is apparently independent of meiotic chromosome pairing. The *mei-2* mutant of *Neurospora*, deficient in meiotic recombination and chromosome pairing, is capable of RIP (FOSS and SELKER 1991).

It is not clear whether DNA methylation is part of the RIP mechanism. Two models for the observed C:G to T:A transition mutations are most favored. In one model, RIP causes methylation of cytosines, followed by enzymatic deamination of some or all of the 5-methylcytosines, to give thymines. In a second model, cytosines are deaminated to become uracils, which are replaced by thymines during replication. The first model requires a failure of G:T mismatch repair, and the second model requires a failure of uracil DNA-glycosylase. Neither model can be ruled out, but the following observations may be relevant. First, a mutant strain of *Neurospora* containing no detectable methylation is capable of RIP (H. Foss and E. Selker, unpublished). Second, some products of RIP are not methylated (SELKER and GARRETT 1988; Singer and Selker, in preparation). Third, enzymatic deamination of cytosine has been demonstrated in vitro with two bacterial methyltransferases, M.Hpa II (SHEN et al. 1992) and M. EcoRII (WYSZYNSKI et al. 1994), which supports the idea that a DNA methyltransferase may also function as a DNA-cytosine deaminase (SELKER 1990b).

2.6 Recurrence of, and Resistance to, RIP

Duplications that have undergone RIP become resistant, but not immune, to RIP (CAMBARERI et al. 1991; FOSS et al 1991). In repeated backcrosses of a strain containing a duplication, RIP recurred but with decreased efficiency. The observation that elements of a duplication do not contain identical mutations (CAMBARERI et al. 1989; GRAYBURN and SELKER 1989; SINGER and SELKER, in preparation) suggested that sequence divergence might be the primary basis for resistance to RIP. Support for this idea came from experiments with altered sequences that had acquired resistance to RIP through multiple sexual cycles. When one such sequence was cloned and introduced into wild-type *Neurospora*, it underwent RIP at high frequency in multiple copy transformants (CAMBARERI et al. 1991). In contrast, a single copy of the cloned sequence did not interact with the native unmutated sequence to trigger RIP. The dependence of RIP on a high degree of sequence similarity is additional evidence that efficient DNA pairing is a prerequisite for efficient RIP.

2.7 Possible Functions of RIP

Gene inactivation by RIP is irreversible, in contrast to gene silencing by epigenetic means. *Ascobolus immersus*, another ascomycete, and *Coprinus cinereus*, a basidiomycete, inactivate genes reversibly by methylation-induced premeiotically (MIP, BARRY et al. 1993; FAUGERON et al. 1990; FREEDMAN and PUKKILA 1993). This phenomenon is very similar to RIP, except that mutations do not occur. As in RIP, duplicated sequences are modified by MIP in a pairwise fashion during the sexual cycle (FAUGERON et al. 1990). RIP and MIP may be defense mechanisms aimed at selfish DNAs such as transposable elements. Indeed, active versions of the transposon, Tad, have been found in only one wild-type strain of *Neurospora crassa*, but relics of Tad exist in all species and strains of *Neurospora* tested (KINSEY 1989; KINSEY and HELBER 1989; KINSEY et al. 1994). As mentioned in Sect. 2.6, RIP causes duplicate sequences to diverge. For example, the elements of one natural tandem duplication diverged from each other by ≈14% due to RIP (GRAYBURN and SELKER 1989). Relative to its native counterpart, a sequence altered by RIP can show substantially greater divergence (CAMBARERI et al. 1991). An advantage of rapid divergence of repeated sequences may be prevention of gross chromosomal rearrangements, such as would result from recombination between unlinked duplications.

3 Premeiotic Deletion/Recombination

Repeat-induced point mutation is not the only phenomenon in *Neurospora* affecting repetitive sequences prior to meiosis. Direct repeats are prone to premeiotic deletion of one copy of the duplicated sequence and of the intervening segment (SELKER et al. 1987; BOWRING and CATCHESIDE 1993). The frequency of premeiotic deletion can be greater than 50%. In multiple backcrosses of a strain containing a direct tandem duplication, the frequency of premeiotic deletion diminished faster than the frequency of RIP, suggesting that premeiotic recombination requires greater homology than does RIP (CAMBARERI et al. 1991). Recent findings suggest that these two processes may also differ in their dependence on length of homologous sequences. A direct duplication of a ≈2-kb fragment separated by a unique sequence of similar size showed efficient RIP without high-frequency deletion (Irelan et al. 1994).

The nucleolar organizing region (NOR) also is subject to premeiotic recombination (BUTLER and METZENBERG 1989; BUTLER and METZENBERG 1990). The NOR consists of an array of ≈100–200 tandem copies of a ≈9-kb unit encoding the three largest ribosomal RNAs. Expansions and contractions of the NOR are common, and up to ≈50% of the repeat units can be lost in a single sexual cycle. It is not clear whether these fluctuations in the NOR are related to the premeiotic recombination observed with other duplications.

Both RIP and premeiotic recombination affect pairs of homologous DNA sequences. The discovery of premeiotic deletion raised the possibility that RIP could be mechanistically related to recombination (SELKER et al. 1987; CAMBARERI et al. 1991; IRELAN et al., in preparation). Unequal sister-chromatid exchange is probably not responsible for the deletions, since direct duplications do not generate triplications (SELKER et al. 1987; IRELAN et al. 1994). One model accounts for deletion in a direct linked duplication by intrachromosomal crossover ("loop-out"). To test this model, an inverted duplication flanking a unique sequence was examined for intrachromosomal crossovers. Inversion of the intervening sequence was detectable, but its frequency was lower than the frequency of deletion of this sequence when flanked by direct repeats (IRELAN et al. 1994). This observation suggests that reciprocal crossovers are not responsible for the high frequency of deletion observed in direct repeats. Because RIP also affects unlinked duplications at a high frequency, the possibility that ectopic recombination is associated with RIP was explored. In strains with two copies of the *mtr* gene, one native and one ectopic, RIP occurred at high frequency, but was not accompanied by gene conversion or crossover (IRELAN et al. 1994). The lack of a correlation between RIP and ectopic recombination suggests that either these processes are mechanistically unrelated or that RIP works against recombination between unlinked repeats.

4 DNA Methylation

4.1 Functions of Methylation

The control and function(s) of DNA methylation are not clearly understood in *Neurospora*, or for that matter, in any eukaryote. DNA methylation is generally correlated with gene inactivity (reviewed in JOST and SALUZ 1993). Numerous studies demonstrate that methylation of promoter sequences can repress transcription in higher eukaryotes, although the extent to which organisms exploit this tool is unclear. Methylation can affect transcription in *Neurospora* as well. Preliminary results involving genes altered by RIP indicate that some methylated *Neurospora* sequences can be transcriptionally reactivated by preventing DNA methylation (Singer and Selker, unpublished; Rountree, Irelan and Selker, unpublished; see also Sect. 5 below).

Neurospora, unlike the mouse (LI et al. 1992), appears to have no absolute requirement for DNA methylation. A mutant of *Neurospora* having no detectable cytosine methylation has been isolated (FOSS et al. 1993).

4.2 Distribution and Control of Methylation

In higher eukaryotes, 5-methylcytosine is primarily located in symmetrical sequences such as the dinucleotide CpG. Such methylation is hypothesized to be stably propagated by a "maintenance methylase" that acts upon hemimethylated sites in newly synthesized DNA (GRUENBAUM et al. 1981; HOLLIDAY and PUGH 1975; RIGGS 1975). In *Neurospora*, specific chromosomal regions are faithfully methylated, but methylated cytosines are not limited to symmetrical sites (SELKER and STEVENS 1985). Genomic sequencing revealed no preference of the methylase for any oligonucleotide sequence (SELKER et al. 1993a). Furthermore, methylation is heterogeneous within clonal cultures of *Neurospora*. These observations conflict with the maintenance methylase model and suggest that *Neurospora* may perpetually reestablish methylation in specific regions, but lacks a system to propagate specific patterns of cytosine methylation (SELKER 1990a). Direct attempts to detect maintenance of methylation patterns imposed on transforming DNA have not revealed evidence of maintenance methylation of arbitrary sequences in *N. crassa* (Lommel, Cambareri, Miao, Margolin and Selker unpublished). This distinguishes *Neurospora* from *Ascobolus* since the existence of MIP implies that *Ascobolus* is capable of maintenance methylation (RHOUNIM et al. 1992; SELKER 1993).

As mentioned above, cytosine methylation is often associated with RIP. The best characterized methylated DNA sequences are : (1) a naturally occurring relic of RIP, the zeta-eta (ζ–η) region (SELKER and STEVENS 1985) and (2) an allele of the *am* gene, $am^{RIP\text{-}8M}$, that was obtained by induction of RIP in the laboratory (SELKER et al. 1993a). Both of these sequences, but not their unaltered progenitors, cause methylation de novo (SELKER et al. 1987b; SELKER and STEVENS 1985; Singer and Selker, in preparation). Several types of models can be imagined for how point mutations cause normally unmethylated sequences to become methylated. In one class of models, the mutations create *cis*-acting oligonucleotide sequences that invite methylation. In another class of models, a higher order (non-sequence-specific) change in the structure or property of a region triggers DNA methylation. It is possible, for example, that the increased A + T content and/or increased density of TpA dinucleotides resulting from RIP somehow leads to methylation.

The tandemly repeated rDNA represents the only example of an endogenous methylated region in *Neurospora* that is not obviously a relic of RIP (PERKINS et al. 1986; RUSSELL et al. 1985). It is not clear whether methylation of the rDNA is due to its sequence, its nuclear location, its high copy number, or some other unknown factor. Another possibility is that RIP occurs in a fraction of the repeats, and that the contractions and expansions of the NOR reflect repair or elimination of mutated repeat units.

Transforming DNA, including foreign DNA sequences and some *Neurospora* sequences, can be methylated in the absence of RIP (BULL and WOOTTON 1984; ORBACH et al. 1988; SELKER et al. 1987b). Methylation of bacterial vector sequences in asexually propagated transformants appears to be position dependent, because it occurs in some single or multicopy transformants but not in others (SELKER et al. 1987b). While duplication of sequences is not sufficient to trigger methylation in asexual cells of *Neurospora* (transformants with multiple copies of unmethylated transforming DNA are common, SELKER and GARRETT 1988; MIAO et al. 1994), copy number may play a role in the methylation of trans-genes (ORBACH et al. 1988; PANDIT and RUSSO 1992; ROMANO and MACINO 1992). In one case, loss of methylation of bacterial vector sequences in a two-copy transformant was associated with loss of one of two copies of the transforming DNA (SELKER et al. 1993b).

5 Quelling/Asexual Inactivation of Genes

5.1 Observations on Asexual Inactivation

Gene duplications are not normally inactivated in the vegetative phase of the *Neurospora* life cycle. Several duplications that underwent RIP at high frequency in the sexual cycle were found to be stable in vegetative cells (Garrett, Irelan, Kinsey and Selker, unpublished). Nevertheless, asexual inactivation of transgenes and their endogenous homologs has been reported in *Neurospora* (PANDIT and RUSSO 1992; ROMANO and MACINO 1992). Romano and Macino observed inactivation of endogenous genes when they transformed with homologous sequences, and termed the phenomenon "quelling". The *al-1* and *al-3* (albino) genes were quelled by fragments of the respective genes, including fragments not containing promoter sequences. Reduced steady-state concentrations of the transcript were observed in affected transformants, suggesting that quelling acts at the level of transcription or transcript stability. Spontaneous reversion of one transformant to the wild-type phenotype was associated with reduction in copy number of the transforming DNA (ROMANO and MACINO 1992). PANDIT and RUSSO reported inactivation of a foreign gene (*hph*) in some transformants with a copy number greater than one, but no inactivation in single-copy transformants. Phenotypic reactivation was induced by 5-azacytidine in some cases, suggesting that cytosine methylation may play a role in inactivation in these cases. High frequencies of reversion reported in both studies indicate that quelling does not involve mutations. Spontaneous reactivation was reversible in one study (PANDIT and RUSSO 1992), but was unidirectional in the other (ROMANO and MACINO 1992). Thus, *Neurospora* has at least one asexual mechanism for epigenetic inactivation of repeated genes.

5.2 Possible Mechanisms of Asexual Inactivation

Several possibilities for the mechanism of asexual inactivation can be imagined. Gene expression could be inhibited by: (1) DNA methylation, (2) an alternate chromatin structure, (3) titration of regulatory factors, (4) altered nuclear location of repeated sequences, or (5) antisense RNA or other post-transcriptional means. DNA methylation is an attractive candidate, but it does not appear to be strictly correlated with inactivation. If methylation is involved, it is not clear how methylation is directed specifically to repeated sequences. Models involving other epigenetic factors, such as chromatin structure, are attractive but difficult to test. ROMANO and MACINO (1992) concluded that antisense RNA could not be responsible for quelling because inactivation occurred at high frequency when they transformed with gene fragments lacking a fungal promoter. Similarly, titration of transcription factors would not be predicted to occur when a promoter is not included in the transforming DNA.

At this point, no single model unites all of the repetitive-gene silencing phenomena observed in fungi and plants. RIP and MIP are most likely initiated by a DNA-DNA interaction between a pair of homologous sequences, but there is no strong evidence that pairing is involved in asexual phenomena such as quelling, co-suppression, and *trans*-inactivation (for more discussion of this point see Rossignol and Faugeron 1994). It will be exciting to elucidate the silencing mechanisms and discover how widely they conserved.

Acknowledgements. We thank Jette Foss, Jeff Irelan, Garth Patterson, and Mike Rountree for helpful comments on the manuscript and J.F., J.I., M.R., Ann Hagemann, and Jack Kinsey for communication of results prior to publication. The work from our laboratory was supported by NIH grant GM 35690 and NSF grant DCB 8718163 and was done during the tenure of an Established Investigatorship of the American Heart Association (E.U.S.).

References

Barry C, Faugeron G, Rossignol J-L (1993) Methylation induced premeiotically in Ascobolus: coextension with DNA repeat lengths and effect on transcript elongation. Proc Natl Acad Sci USA 90: 4557–4561

Bell PD, Dunlap JC, Loros JJ (1992) The Neurospora circadian clock-controlled gene, ccg-2, is allelic to eas and encodes a fungal hydrophobin required for formation of the conidial rodlet layer. Genes Dev 6: 2382–2394.

Bowring FJ, Catcheside DEA (1993) The effect of rec-2 on repeat-induced point-mutation (RIP) and recombination events that excise DNA sequence duplications at the his-3 locus in Neurospora crassa. Curr Genet 23: 496–500

Bull JH, Wootton JC (1984) Heavily methylated amplified DNA in transformants of Neurospora crassa. Nature 310: 701–704

Butler DK, Metzenberg RL (1989) Premeiotic change of nucleolus organizer size in Neurospora. Genetic 122: 783–791

Butler DK, Metzenberg RL (1990) Expansion and contraction of the nucleus organizer region of Neurospora: changes originate in both proximal and distal segments. Genetics 126: 325–333

Cambareri EB, Jensen BC, Schabtach E, Selker EU (1989) Repeat-induced G-C to A-T mutations in Neurospora. Science 244: 1571–1575

Cambareri EB, Singer MJ, Selker EU (1991) Recurrence of repeat-induced point mutation (RIP) in Neurospora crassa. Genetics 127: 699–710

Connerton I (1990) Premeiotic disruption of the Neurospora crassa malate synthase gene by native and divergent DNAs. Mol Gen Genet 223: 319–323

Faugeron G, Rhounim L, Rossignol J-L (1990) How does the cell count the number of ectopic copies of a gene in the premeiotic inactivation process acting in Ascobolus immersus? Genetics 124: 585–591

Fincham JR (1990) Generation of new functional mutant alleles by premeiotic disruption of the Neurospora crassa am gene. Curr Genet 18: 441–445

Fincham JRS, Connerton IF, Notarianni E, Harrington K (1989) Premeiotic disruption of duplicated and triplicated copies of the Neurospora crassa am (glutamate dehydrogenase) gene. Curr Genet 15: 327–334

Foss HM, Selker EU (1991) Efficient DNA pairing in a Neurospora mutant defective in chromosome pairing. Mol Gen Genet 231: 49–52

Foss EJ, Garrett PW, Selker EU (1991) Specificity of repeat induced point mutation (RIP) in Neurospora: sensitivity of non-Neurospora sequences, a natural diverged tandem duplication, and unique DNA adjacent to a duplicated region. Genetics 127: 711–717

Foss HM, Roberts CJ, Claeys KM, Selker EU (1993) Abnormal chromosome behavior in Neurospora mutants defective in DNA methylation. Science 262: 1737–1741

Freedman T, Pukkila PJ (1993) De novo methylation of repeated sequences in Coprinus cinereus. Genetics 135: 357–366

Glass NL, Lee L (1992) Isolation of Neurospora crassa A mating type mutants by repeat induced point (RIP) mutation. Genetics 132: 125–133

Grayburn WS, Selker EU (1989) A natural case of RIP: degeneration of DNA sequence in an ancestral tandem duplication. Mol Cell Biol 9: 4416–4421

Gruenbaum Y, Naveh-Many T, Cedar H, Razin A (1981) Sequence specificity of methylation in higher plant DNA. Nature 292: 860–862

Harkness TAA, Metzenberg RL, Schneider H, Lill R, Neupert W, Nargang FE (1994) Inactivation of the Neurospora crassa gene encoding the mitochondrial protein import receptor MOM19 by the technique of "sheltered RIP". Genetics 136: 107–118

Holliday R, Pugh JE (1975) DNA modification mechanisms and gene activity during development. Science 187: 226–232

Irelan JT, Hagemann AT, Selker EU (1994) High frequency repeat-induced point mutation (RIP) is not associated with efficient recombination in Neurospora. Genetics 138: 1093–1103

Jarai G, Marzluf GA (1991) Generation of new mutants of nmr, the negative- acting nitrogen regulatory gene of Neurospora crassa, by repeat induced mutation. Curr Genet 20: 284–288

Jost JP, Saluz HP (eds) (1993) DNA methylation: molecular biology and biological significance. Birkhäuser, Boston

Kinsey JA (1989) Restricted distribution of the Tad transposon in strains of Neurospora. Curr Genet 15: 271–275

Kinsey JA, Helber J (1989) Isolation of a transposable element from Neurospora crassa. Proc Natl Acad Sci USA 86: 1929–1933

Kinsey JA, Garrett-Engele PW, Cambareri EB, Selker EU (1994) The Neurospora transposon Tad is sensitive to repeat-induced point mutation (RIP). Genetics 138: 657–664

Li E, Bestor TH, Jaenisch R (1992) Targeted mutation of the DNA methyltransferase gene results in embryonic lethality. Cell 69: 915–926

Marathe S, Connerton IF, Fincham JRS (1990) Duplication-induced mutation of a new Neurospora gene required for acetate utilization: properties of the mutant and predicted amino acid sequence of the protein product. Mol Cell Biol 10: 2638–2644

Matzke M, Matzke AJM (1993) Genomic imprinting in plants: parental effects and trans-inactivation phenomena. Annu Rev Plant Physiol Plant Mol Biol 44: 53–76

Miao VPW, Singer MJ, Rountree MR, Selker EU (1994) Identification of signals for de novo methylation in Neurospora crassa by targeted gene replacement. Mol Cell Biol 14: 7059–7067

Nelson MA, Metzenberg RL (1992) Sexual development genes of Neurospora crassa. Genetics 132: 149–162

Orbach MJ, Schneider WP, Yanofsky C (1988) Cloning of methylated transforming DNA from Neurospora crassa in Escherichia coli. Mol Cell Biol 8: 2211–2213

Pandit NN, Russo VEA (1992) Reversible inactivation of a foreign gene, hph, during the asexual cycle in Neurospora crassa transformants. Mol Gen Genet 234: 412–422

Perkins DD, Barry ED (1977) The cytogenetics of Neurospora. Adv Genet 19: 133–285

Perkins DD, Metzenberg RL, Raju NB, Selker EU, Barry EG (1986) Reversal of a Neurospora translocation by crossing over involving displaced rDNA, and methylation of the rDNA segments that result from recombination. Genetics 114: 791–817

Rhounim L, Rossignol J-L, Faugeron G (1992) Epimutation of repeated genes in Ascobolus immersus. EMBO J 11: 4451–4457

Riggs AD (1975) X-inactivation, differentiation and DNA methylation. Cytogenet. Cell Genet 14: 9–25

Romano N, Macino G (1992) Quelling: transient inactivation of gene expression in Neurospora crassa by transformation with homologous sequences. Mol Microbiol 6: 3343–3353

Rossignol J-L, Faugeron G (1994) Gene inactivation triggered by recognition between DNA repeats. Experientia 50: 307–317.

Russell PJ, Rodland KD, Cutler JE, Rachlin EM, McClosley JA (1985) DNA methylation in Neurospora: chromatographic and isischizomer evidence for changes during development. In: Timberlake W (eds) Molecular genetics of filamentous fungi. Liss, New York, pp 321–332

Selker EU (1990a) DNA methylation and chromatin structure: a view from below. Trends Biochem Sci 15: 103–107.

Selker EU (1990b) Premeiotic instability of repeated sequences in Neurospora crassa. Annu Rev Genet 24: 579–613

Selker EU (1993) Control of DNA methylation in fungi. In: Jost JP, Saluz HP, (eds) DNA methylation: molecular biology and biological significance. Birkhäuser, Boston, pp 212–217

Selker EU, Garrett PW (1988) DNA sequence duplications trigger gene inactivation in Neurospora crassa. Proc Natl Acad Sci USA 85: 6870–6874

Selker EU, Cambareri EB, Jensen BC, Haack KR (1987a) Rearrangement of duplicated DNA in specialized cells of Neurospora. Cell 51: 741–752

Selker EU, Stevens JN (1985) DNA methylation at asymmetric sites is associated with numerous transition mutations. Proc Natl Acad Sci USA 82: 8114–8118

Selker EU, Jensen BC, Richardson GA (1987b) A portable signal causing faithful DNA methylation de novo in Neurospora crassa. Science 238: 48–53

Selker EU, Cambareri EB, Garrett PW, Jensen BC, Haack KR, Foss E, Turpen C, Singer M, Kinsey J (1989) Use of RIP to inactivate genes in Neurospora crassa. Fungal Genet Newslett 36: 76–77

Selker EU, Fritz DY, Singer MJ (1993a) Dense non-symmetrical DNA methylation resulting from repeat-induced point mutation (RIP) in Neurospora. Science 262: 1724–1728

Selker EU, Richardson GA. Garrett-Engele PW, Singer MJ, Miao V (1993b) Dissection of the signal for DNA methylation in the ζ–η region of Neurospora. Cold Spring Harb Symp Quant Biol 58: 323–329

Shen J-C, Rideout WM III, Jones PA (1992) High frequency mutagenesis by a DNA methyltransferase. Cell 71: 1073–1080

Stadler D, Macleod H, Dillon D (1991) Spontaneous mutation at the mtr locus of Neurospora: the spectrum of mutant types. Genetics 129: 39–45

Wyszynski M, Gabbara S, Bhagwat AS (1994) Cytosine deaminations catalyzed by DNA cytosine methyltransferases are unlikely to be the major cause of mutational hot spots at sites of cytosine methylation in Escherichia coli. Proc Natl Acad Sci USA 91: 1574–1578

MIP: An Epigenetic Gene Silencing Process in *Ascobolus immersus*

J.-L. Rossignol and G. Faugeron

1 Introduction

The filamentous fungus *Ascobolus immersus* is a haploid and heterothallic Ascomycete belonging to the class Discomycetes ("cup fungus"). *Ascobolus* was introduced as material for genetic studies by Georges Rizet (Lissouba et al. 1962). It has been used for years as a model organism for investigating meiotic recombination. In this species, each ascus is composed of eight haploid spores (four pairs), and each pair of spores is issued through a mitotic division from one of the four meiotic products arising from an individual meiosis. Wild-type spores display a dark-brown pigmentation. Spore color mutants exhibiting white ascospores can be screened, using a binocular microscope. In crosses between wild-type and mutant strains, mendelian segregations yield four wild-type (dark-brown) and four mutant (white) spores. Asci showing departures from mendelian segregations are easily scored. They reflect the occurrence of gene conversion events resulting from the process of meiotic recombination. One spore color gene, named *b2*, was extensively used for studying meiotic recombination. The genetic study of segregation and recombination patterns in crosses heterozygous

---

Institut de Génétique et Microbiologie, Bâtiment 400, CNRS: URA 1354, Université Paris-Sud, 91405 Orsay cedex, France

for one or several allelic mutations in *b2* provided valuable information about the formation of recombination intermediates (reviewed in Rossignol et al. 1988; Nicolas and Rossignol 1989).

In order to study genetic recombination at a molecular level, we have developed genetic transformation tools. The initial goals were on the one hand to use genetic transformation to clone the *b2* spore color gene by complementing *b2* mutant strains, and on the other hand to allow the construction of genetically engineered strains via gene replacement, using in vitro modified transgenes. Several methods of gene replacement were actually set up (Goyon and Faugeron 1989), whereas the cloning of *b2* had to be achieved by another procedure using genomic subtraction (Colot and Rossignol, manuscript in preparation). The analysis of transformant strains led Faugeron et al. (1989) to an unexpected observation: transformants resulting from either homologous or nonhomologous integration were highly unstable through sexual reproduction. The loss of the transformant phenotype was shown to result from a premeiotic event leading to the inactivation of the transgene (Goyon and Faugeron 1989). This event occurred only when two identical DNA segments (e.g., the transgenic copy and the resident one) were present in the haploid transformant strain. DNA analysis showed that transgene inactivation was associated with methylation of the cytosine residues belonging to the duplicated segment. This process was later termed MIP (standing for methylation induced premeiotically). We shall summarize here the main features of MIP.

2 Methylation Induced Premeiotically (MIP) Is Triggered by the Presence of DNA Repeats

In the first experiments, a *met2* mutant strain was transformed by plasmids harboring the wild-type *met2* allele. This led to four types of transformants (Fig. 1). Types i and ii had integrated transgenic copies into ectopic positions, via nonhomologous integrative recombination; type i had integrated only one transgenic copy (Faugeron et al. 1989) whereas type ii had integrated several copies clustered at a single site (Rhounim et al. 1994). Type iii had integrated one or several transgenic copies via homologous recombination at the resident *met2* locus and thus bore tandem repeats of *met2* (Goyon and Faugeron 1989). Type iv resulted from the substitution of the resident mutated allele by the transgenic wild-type *met2* allele (Faugeron et al. 1989). All these transformants displayed a stable Met$^+$ phenotype through vegetative propagation of the mycelium, but they were not all stable through sexual reproduction. Almost 100% of the progeny of type ii and iii transformants (Goyon and Faugeron 1989; Rhounim et al. 1994) and about 50% of the progeny of type i transformants (Faugeron et al. 1989) lost the Met$^+$ transformant phenotype through sexual reproduction. Only type iv transformants remained fully stable (Faugeron et al. 1989). Southern analysis of DNA from

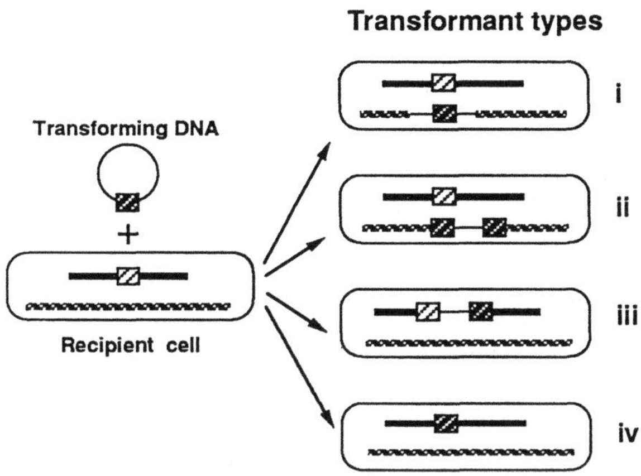

Fig. 1. Different types of transformants obtained using a transgene (*thick-hatched box*) from *Ascobolus*. The homologous resident gene in the recipient cell (*thin-hatched box*) can correspond, according to the experiments, either to the mutant or to the wild-type allele. Vector DNA is represented by a *thin line*; two distinct chromosomes in the recipient cell are represented (*thick black line*, harboring the resident gene, and *thick hatched line*). Type i: integration of a single transgenic copy at an ectopic site (unlinked to the resident gene). .Type ii: integration of several transgenic copies at a single ectopic site. Type iii: integration of one transgenic copy (or several such copies) via homologous recombination at the resident locus, leading to a direct tandem repeat of the transgenic and resident copies. Type iv: substitution of the resident copy by the transgenic one

the auxotroph derivatives of type i, ii and iii transformants showed that the transgene had not been excised: the loss of the transformant phenotype thus resulted from transgene inactivation.

The observation that instability through sexual reproduction occurred with all type i, ii, and iii transformants that contained several copies of *met2* (the resident copy, plus at least one transgenic copy) but not with type iv transformants that contained only one copy of it suggested that the MIP inactivation process was dependent on the presence of more than one copy of the target gene in the same haploid parent. Further observations clearly showed that this was the only condition for triggering MIP. Transformation experiments were set up, using the *amdS* gene from *Aspergillus nidulans*, which is absent from the wild-type *Ascobolus* strain. Two types of transformants, selected for their ability to use acetamide as nitrogen source (Amd$^+$ phenotype), were obtained. Type 1 transformants each contained a single transgenic copy (Fig. 2) integrated at different genomic sites according to the transformant (Faugeron et al. 1990). Type 2 transformants each contained several transgenic copies clustered at a single genomic site (Rhounim et al. 1994). In crosses with the wild-type strain, type 1 transformants appeared fully stable, transmitting the Amd$^+$ phenotype to the whole transformant progeny (Faugeron et al. 1990). In contrast, crosses of type 2 transformants with wild type always gave a progeny which had completely lost the Amd$^+$ phenotype (Rhounim et al. 1994). Faugeron et al. (1990) intercrossed type 1 transformants (for example 1a and 1b, see Fig. 2). In such crosses, where

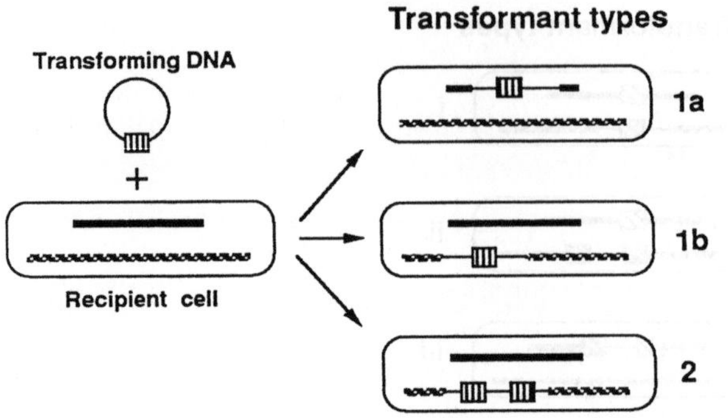

Fig. 2. Different types of transformants obtained using a foreign transgene (*hatched box*) which has no equivalent in the recipient cell. Vector DNA is represented by a *thin line*; two distinct chromosomes in the recipient cell are represented (*thick black line* and *thick hatched line*). Type 1 transformants have integrated one transgenic copy; in types 1a and 1b, the integration occurred at different sites. Type 2 transformants have integrated several copies clustered at a single site

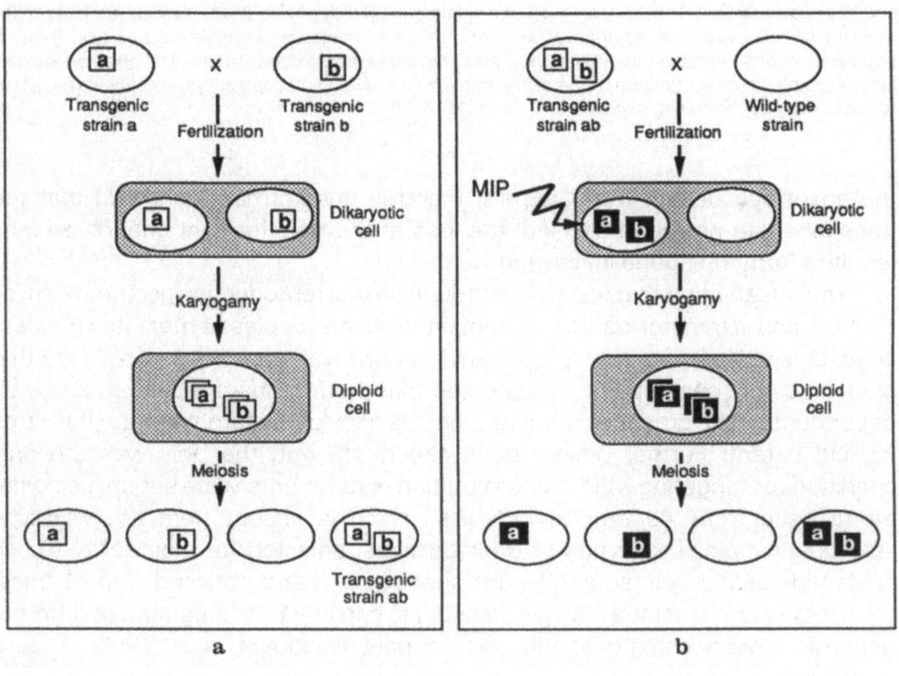

Fig. 3 a, b. Timing and conditions of MIP triggering. Panel **a** represents a cross between two haploid strains, each carrying a single copy (copy *a* or copy *b*) of a foreign transgene integrated at two different genomic sites, as do transformants 1a and 1b of Fig. 2. In panel **b**, and *a* and *b* copies are in the same parental strain. Fertilization in *Ascobolus* results in formation of dikaryotic cells containing each of the two parental haploid nuclei. These cells will undergo several mitotic divisions, followed by karyogamy. In the resulting diploid nuclei which enter meiosis, each copy has undergone premeiotic DNA replication (represented by the *a* and *b duplicated boxes*). In panel **b**, transgene inactivation resulting from MIP is represented by *blackened boxes*. In each panel, the four possible meiotic products are shown

each parent harbored one copy of *amdS* (inserted at different genomic sites), MIP never occurred. In the progeny of these crosses (Fig. 3a), they isolated Amd$^+$ strains which had inherited both transgenic copies, one from each parent, as a result of meiotic segregation. These strains that contained two unlinked *amdS* copies were crossed with the wild-type strain (Fig. 3b). In the progeny of these crosses, some (about 50%) of the asci displayed inactivation of *amdS*. Inactivation appeared in all the products of individual meiosis having inherited the *amdS* transgene, indicating that both copies had been inactivated simultaneously. This showed that MIP occurs before meiosis. Otherwise, crosses where each of the two *amdS* copies is brought by one parent and crosses where both *amdS* copies are brought by the same parent should have produced the same result, since they give identical diploid cells after karyogamy. Thus, MIP is initiated at some time between fertilization and karyogamy, and its triggering needs the presence of DNA repeats (at least two copies of the same DNA segment) in the same haploid parent.

3 Stability and Reversibility of MIP Silencing

One transformant harboring two unlinked wild-type copies of *met2* (the resident copy and the transgenic copy integrated at an ectopic site) was further obtained. When this transformant was put into a cross, both *met2* copies became simultaneously inactivated by MIP (Rhounim et al. 1992), as was the case with *amdS*. Strains having inherited only one copy of *met2* (the inactivated resident wild-type copy), as a result of meiotic segregation, were isolated from the progeny and used to study the stability of inactivation (Rhounim et al. 1992). It was shown that, in the absence of selection, inactivation was faithfully maintained through mitotic divisions. Although stable, inactivation can be reversed. All the strains tested (several hundred), harboring either one or two copies of the inactivated *met2* gene, were shown to revert when put in selective conditions. This spontaneous reversion was usually detected 5–30 days after transfer onto selective medium. In this case, reversion probably occurred first in one nucleus among the thousands of nuclei which were present in the piece of mycelium placed on selective medium. Reversion occurred faster in the presence of the demethylating agent 5-azacytidine. This shorter delay is likely to reflect an increased probability per nucleus to revert, consistent with the involvement of cytosine methylation in inactivation (see below). These observations suggest that inactivation corresponds to a reversible epigenetic silencing process. The fact that inactivation persisted in individual meiotic products that had inherited only one of the two copies which had been subject to MIP indicates that two copies are not needed for maintaining silencing, in contrast to the requirement of two copies for triggering MIP. Despite the fact that they are all able to revert, the silenced strains are remarkably stable, even after sexual reproduction, since it was shown that inactivation was faithfully maintained through five successive sexual cycles.

Interestingly, in the progeny of crosses between one parent harboring only the inactivated resident copy and one parent harboring an active wild-type resident copy, the inactivated state segregated as a mutant allele (i.e., faithful mendelian segregations were observed).

4 Intensity and Extent of Methylation Resulting from MIP

The first indication that MIP silencing was associated with cytosine methylation was obtained by GOYON and FAUGERON (1989), who studied transformants harboring a tandem duplication of a 5.7-kb fragment, obtained by homologous integration of the transforming DNA. This tandem was constituted of the *met2* mutant (resident) allele and the wild-type *met2* (transgenic) allele. Southern analysis of the DNA from Met$^-$ derivatives indicated that both the resident and the transgenic *met2* copies had undergone methylation (restriction enzymes sensitive to cytosine methylation did not cut their recognition sites located within the duplicated segment).

A further study was carried out upon transformants having two wild-type copies of *met2* duplicated either in tandem array, following homologous integration, or in unlinked positions, following nonhomologous integration (RHOUNIM et al. 1992). Whatever the duplication, MIP silencing always involved both *met2* copies. Southern analysis using pairs of restriction enzymes that recognize the same restriction site, but that are either sensitive or insensitive to cytosine methylation, showed that restriction sites located within the duplicated segments had undergone cytosine methylation and were never mutated. The absence of mutation was further demonstrated by DNA sequence analysis which failed to show any change in the base sequence of the silenced genes. This important finding clearly distinguishes MIP from RIP (repeat-induced point mutation), a related process acting in the filamentous Ascomycete *Neurospora crassa*, which not only leads to cytosine methylation, as does MIP, but also to C to T transition mutations (SELKER 1990a; see also the chapter by M.J. Singer and E.U. Selker, this volume). The source of the differences between MIP and RIP was discussed (ROSSIGNOL and FAUGERON 1994). Because of the similarity between the two processes, it would have been possible that MIP was in fact a deficient RIP (i.e., in the laboratory *Ascobolus* strain studied, a mutation would have changed the RIP process into MIP, allowing methylation only, and preventing point mutations). This possibility was checked in another *Ascobolus* strain recently isolated from the wild. MIP (and no RIP) was observed in this strain (Faugeron, Gregoire and Rossignol, unpublished results), confirming that MIP is indeed a biologically relevant process not restricted to laboratory strains.

Transformant strains in which *met2* fragments of different sizes (1.2–5.7 kb) had been integrated at positions unlinked to the resident wild-type *met2* copy were studied with respect to MIP by Barry et al. (1993). The eight resulting duplications (which encompassed distinct parts of the *met2* gene) were all able to trigger methylation of the resident gene during sexual reproduction. Southern analysis showed that the extent of methylation depended on the size of the duplication and was coextensive with it, possibly with lower intensity close to the borders. Barry, Faugeron and Rossignol (unpublished results) duplicated smaller and smaller fragments of the *met2* open reading frame (ORF). When repeats were unlinked, the frequency of MIP triggering decreased with the size of the duplication. The smallest duplication tested was 630 bp long. It was subject to MIP (monitored by the methylation of restriction sites in the resident gene) with a frequency of 3%. They also tested tandem duplications of decreasing sizes and showed that a 457-bp duplication was still able to efficiently trigger methylation of the duplicated portion. In contrast, no methylation could be detected when the size of the tandem duplication dropped to 317 bp.

Goyon et al. (1994) used the bisulfite genomic sequencing method (Frommer et al. 1992) to determine the methylation status of every cytosine residue in a strain in which the entire ORF of the *met2* gene had undergone MIP. Several individual molecules all issued from the replication of a single molecule initially subject to MIP were sequenced. In each molecule, the methylation extended over almost the whole length of the previously duplicated segment. The extent of methylation was precisely delimited and constant in each of the molecules, leaving unmethylated a close to 100-bp region located within the duplicated segment next to each end. In none of the molecules did the methylation resulting from MIP extend beyond the ends. Although all the DNA molecules were not methylated with the same intensity, all cytosine residues in the methylated portion could be methylated (the level of methylation varying between 43% and 95% of the Cs per DNA molecule, among 39 DNA molecules sequenced). Most of these Cs belonged to nonsymmetrical sequences. This contrasts with the situation in higher eukaryotes, in which most, if not all, methylation is at short symmetrical sequences such as CpG or CpNpG. Only two other cases of dense nonsymmetrical methylation have been reported so far. One concerns methylation in a transgene promoter of *Petunia hybrida* (Meyer et al. 1994). The other case is in *Neurospora*, for genomic segments that have undergone RIP (Selker et al. 1993). However, methylation in *Neurospora* is very likely mediated by a de novo methylase whose recognition signal is constituted by the RIP-induced mutations (Selker 1990b), whereas in *Ascobolus* methylation resulting from MIP has been shown to be perpetuated by a maintenance mechanism (Rhounim et al. 1992) which must involve a sequence-nonspecific maintenance methylase. Methylation in *Ascobolus* was found to be the best maintained at Cs belonging to CpG dinucleotides. Indeed, among 50 Cs belonging to CpGs, 40 were methylated in 100% of the molecules sequenced. This suggests that a CpG-specific methylation activity is superimposed, or coordinated with the sequence-nonspecific maintenance methylation activity.

5 Transcriptional Effects of Methylation Resulting from MIP

The duplication of a subfragment of the *met2* gene is sufficient not only to trigger methylation, but also to silence the resident gene. Indeed, BARRY et al. (1993) showed that silencing of the resident *met2* gene was triggered by duplication of segments corresponding to the region 5′ to the ORF, to only the open reading frame, or to segments beginning 0.87–1.2 kb downstream from the transcription start. The smallest duplication able to trigger silencing was that of the 630-bp segment corresponding to the upstream portion of the *met2* ORF (Barry, Faugeron and Rossignol, unpublished results). However, no gene silencing was detected with the 457-bp duplication, which still triggered methylation.

Northern analysis was performed on silenced strains where methylation involved either the entire *met2* gene or only the region 5′ to the ORF. In both cases transcripts were undetectable. In contrast, a low amount of short transcripts was observed when methylation began downstream from the transcription start (BARRY et al. 1993). Two such cases were further investigated: RT-PCR analysis indicated that these short transcripts were initiated in the same region as in the wild type and had the size expected if methylation were to block transcript elongation. This shows that methylation is responsible for gene silencing by preventing correct transcription of the methylated gene.

6 MIP Affects Both Artificial and Natural DNA Repeats

All the artificially repeated DNA segments (via transformation) tested were shown to undergo MIP during sexual reproduction (FAUGERON et al. 1990; RHOUNIM et al. 1994; Colot, Maloisel and Rossignol, unpublished results). This includes the resident genes *met2* and *b2*, as well as the foreign genes *amdS* from *A. nidulans* and *hph* from *Escherichia coli*. MIP always happens when the DNA repeats are clustered in tandem array (RHOUNIM et al. 1992, 1994), whereas its frequency is variable (between 8% and more than 90% of the premeiotic cells) with unlinked duplications.

Natural DNA repeats are likely to constitute the actual targets of MIP. In order to check this hypothesis, natural DNA repeats were searched for in an *Ascobolus* genomic library, on the basis of the stronger hybridization signal they gave when probed with total DNA (Goyon, Rossignol and Faugeron, unpublished results). Twelve distinct DNA repeats were thus detected. Eight of them clearly showed methylation. In contrast, all the unique DNA sequences tested did not show methylation. Three of the 12 DNA repeats were further investigated. Two of them showed heavy methylation, as would be expected if they were subject to MIP. They have been called *Mars1* and *Mars2* (*Mars* stands for "methylated *Ascobolus* repeated sequence"). *Mars1* is present in more than 50 copies, and corresponds to a LINE-like element. *Mars2* is present in more than ten copies and may also belong to the retrotransposon family. The third DNA repeat studied did not shows methylation. It corresponds to 5S rDNA. The 5S rDNA copies are dispersed in *Ascobolus*. The length of these repeats (120 bp) is smaller than the minimum size required for MIP, determined in experiments using artificial repeats.

rDNA encoding the 28S, 18S, and 5.8S rARN is present in *Ascobolus* in more than 100 tandem repeats, the unit size being almost 9 kb. These repeats also show methylation, but its intensity is much lower than that of usual MIP methylation (Goyon and Faugeron, unpublished results). Further investigation is required to decide whether rDNA is protected against MIP, its methylation resulting from another process, or whether MIP methylation is not efficiently maintained in the rDNA repeats.

7 MIP Involves Homologous DNA-DNA Interactions

The observation that two dispersed copies of the *amdS* gene present in the same haploid parent were always simultaneously silenced when subject to MIP prompted FAUGERON et al. (1990) to check the silencing of three dispersed copies. They observed that MIP never led to the silencing of one copy only, confirming previous results. When it occurred, either two out of the three copies or all three of them were silenced. This observation allowed them to rule out hypotheses according to which all the copies would be silenced at once (as observed in plants when co-suppression occurs: see JORGENSEN 1993, for a review). It constituted an argument favoring the occurrence of a DNA-DNA pairing step allowing the copies to find each other and to become silenced. In this interpretation, the silencing of the third copy would be triggered by two successive pairing events, in which one copy would be involved twice: two copies would be silenced following the first event, the third copy being silenced after the second pairing event. Reconstitution experiments in which one parent strain harbored two silenced copies and one active copy showed that the active copy was frequently subject to MIP, indicating that a methylated inactive copy was thoroughly efficient in triggering MIP of another copy. The hypothesis of a direct DNA-DNA interaction step during the MIP process is strengthened by the observation that the length of the methylated area is dependent upon the length of the duplication (BARRY et al. 1993).

The RIP process acting in *Neurospora*, which is closely related to MIP, is likely to be triggered by the same type of direct DNA-DNA interaction between repeats (SELKER 1990a). The question is open as to whether search for homology in both MIP and RIP is mediated by the same mechanism as in genetic recombination or by different ones (see ROSSIGNOL and FAUGERON 1994, for a discussion).

8 Biological Significance of MIP

That silencing involves pairs of copies appears (up to now) a specific feature of MIP (and RIP as well), as compared to repeat-induced gene silencing phenomena occurring in plants (see Rossignol and Faugeron 1994, for a discussion). These raises the question of the generality of MIP and RIP processes. Several puzzling observations prevent us from providing a simple answer. For example, species closely related to *Neurospora* (i.e., *Sordaria macrospora* and *Podospora anserina*, belonging like *Neurospora* to Pyrenomycetes) do not show RIP (nor MIP), whereas *Ascobolus*, where the RIP-related MIP process operates, belongs to Discomycetes, a fungus class different from Pyrenomycetes. These fungi all belong to the Ascomycotina subdivision. A process, described in *Coprinus cireneus,* which belongs to the Basidiomycotina subdivision, was also called MIP by Freedman and Pukkila (1993). Actually, this process (which also occurred in the dikaryotic phase, after fertilization) led to only slightly methylated repeats and usually did not result in gene silencing. That MIP and RIP appear so scattered among fungus species raises still unsolved questions about their evolutionary origins and their biological roles. If the natural targets of MIP and RIP are natural repeats, then they may be envisioned as processes ensuring genome stability. MIP and RIP might control the mobility of transposons by inactivating them and hence limiting the number of repeats. Cytosine methylation per se might inhibit inter-repeat recombination which generates chromosome rearrangements, just as it inhibits site-specific recombination leading to the maturation of immunoglobulin genes (Goodhart et al. 1993). Then, it is not clear why such processes are observed in some species but not in other, closely related, species. What is the nature of the selection pressures that lead to the losing (or the acquisition) of MIP (or RIP)?

On a more general point, do similar processes targeted at DNA repeats and playing a role in genome stability act in other eukaryotes, particularly plants and animals?

It seems that maize has evolved a mechanism in which cytosine methylation is involved in the control of transposition of mobile elements. The evidence comes from studies on the changes in the activity states of three transposable elements, *Ac*, *Spm*, and *Mu*, for which loss of activity has been clearly correlated with DNA methylation (Fedoroff 1989, for a review). Similarly to MIP methylation, methylation associated with an inactive maize transposable element does not extend into flanking genomic sequences. In addition, the pattern of methylation and the activity state of transposable elements is stably inherited through meiosis, with only a low frequency of reversion to an active state. Whether DNA methylation has a regulatory role in transposition in other plants is not yet known, since only partial data are available. When introduced into tobacco, the *Tam3* element from *Antirrhinum* is able to transpose, and suppression of transposition is associated (but not clearly correlated) with methylation of the inactive elements (Martin et al. 1989). When transferred into tobacco or into *Arabidopsis*, the maize

Ac element displays a low transpositional activity; methylation does not account for such a low activity since, in these plant species, *Ac* remains hypomethylated (Nelsen-Salz and Döring 1990; Keller et al. 1992).

In human, the one million short *Alu* repeats, which result from the retrotransposition of an RNA intermediate, are broadly distributed throughout the genome. They are very rich in CpG dinucleotides (9%), as compared to total human DNA (1%). In somatic tissues these CpGs are extensively methylated (Schmid 1991), while they are highly undermethylated in sperm and their methylation is developmentally regulated (Hellmann-Blumberg et al. 1993). Liu and Schmid (1993) proposed roles for DNA methylation in both *Alu* transcriptional repression and mutational inactivation: first, methylation would limit transcriptional and therefore transpositional competence of *Alu* repeats; second, methylation would inactivate *Alu* repeats through mutations due to very rapid transition of 5-methylcytosine to thymine. This would place the older *Alu* repeats (which are likely to have undergone more inactivating mutations) at a selective disadvantage in the retrotranspositional pathway.

The discovery of RIP in *Neurospora* followed by that of MIP in *Ascobolus* together with the observation that sequence divergence by well-spread point mutations is a potent inhibitor of homologous recombination (Rayssiguier et al. 1989; Waldman and Liskay 1987) prompted Kricker et al. (1992) to hypothesize an active and efficient mechanism possibly used by eukaryotes to maintain genome stability after the amplification of sequences giving identical repeats. The authors proposed and tested a model (by analyzing mammalian and other gene sequences) for a defensive strategy in which inter-repeat recombination is controlled by a germ line process able to identify duplicated sequences by homologous pairing, modify them by CpG methylation, and mutate them by C to T transitions. The model helps to explain the diversity of CpG frequencies in different genes.

These observations in plants and mammals are consistent with the idea that processes, perhaps not totally identical to MIP and RIP, but sharing common features with them, could act in these organisms. Even if processes similar to MIP and RIP are widespread in higher eukaryotes, one should keep in mind that in some organisms, such as yeast and drosophila, cytosine methylation has not been detected, and that the lack or scarcity of methylation may not fit the occurrence of such processes. One may thus speculate that MIP- and RIP-related processes may correspond to strategies developed by living organisms to ensure genome stability along with other processes which still have to be evidenced. For example, the inhibition of transposition by transposition repressors (Finnegan 1992, for a review) means that specific proteins may recognize specific targets in the DNA sequence of mobile elements. In the same way, Wallis et al. (1989) evidenced a gene whose product, a topoisomerase, lowers recombination between δ yeast repeated sequences, and Gottlieb and Easton Esposito (1989) showed that the SIR2 protein inhibited recombination between rDNA repeats in yeast. This suggests that proteins may recognize DNA sequences specific to natural repeats and inhibit inter-repeat recombination by acting as recombination repressors, hence constituting an alternative pathway to RIP or MIP in the stabilization of genomes.

Acknowledgements. We thank Caroline Barry, Vincent Colot, Christophe Goyon, Annie Grégoire, and Laurent Maloisel for communicating unpublished results, and Pierre-Henri Gouyon, Christophe Goyon, and Michel Jacquet for helpful comments on the manuscript.

References

Barry C, Faugeron G, Rossignol J-L (1993) Methylation induced premeiotically in Ascobolus: coextension with DNA repeat lengths and effect on transcript elongation. Proc Natl Acad Sci USA 90: 4557–4561

Faugeron G, Goyon C, Grégoire A (1989) Stable allele replacement and unstable non-homologous integration events during transformation of Ascobolus immersus. Gene 76: 109–119

Faugeron G, Rhounim L, Rossignol J-L (1990) How does the cell count the number of ectopic copies of a gene in the premeiotic inactivation process acting in Ascobolus immersus? Genetics 124: 585–591

Fedoroff NV (1989) About maize transposable elements and development. Cell 56: 181–191

Finnegan DJ (1992) Transposable elements. Curr Opin Genet Dev 2: 861–867

Freedman T, Pukkila PJ (1993) De novo methylation of repeated sequences in Coprinus cinereus. Genetics 135: 357–366

Frommer M, McDonald LE, Millar DS, Collis CM, Watt F, Grigg GW, Molloy PL, Paul CL (1992) A genomic sequencing protocol that yields a positive display of 5-methylcytosine residues in individual DNA strands. Proc Natl Acad Sci USA 89: 1827–1831

Goodhart M, Cavelier P, Doyen N, Kallenbach S, Babinet C, Rougeon F (1993) Methylation status of immunoglobulin κ gene segments correlates with their recombination potential. Eur J Immunol 23: 1789–1795

Gottlieb S, Easton Esposito R (1989) A new role for a yeast transcriptional silencer gene, SIR2, in regulation of recombination in ribosomal DNA. Cell 56: 771–776

Goyon C, Faugeron G (1989) Targeted transformation of Ascobolus immersus and de novo methylation of the resulting duplicated DNA sequences. Mol Cell Biol 9: 2818–2827

Goyon C, Nogueira TIV, Faugeron G (1994) Perpetuation of cytosine methylation in Ascobolus immersus implies a novel type of maintenance methylase. J Mol Biol 240: 42–51

Hellmann-Blumberg U, McCarthy Hintz MF, Gatewood JM, Schmid CW (1993) Developmental differences in methylation of human Alu repeats. Mol Cell Biol 13: 4523–4530

Jorgensen R (1993) The germinal inheritance of epigenetic information in plants. Philos Trans R Soc Lond [Bid] 335: 173–181

Keller J, Lim E, James DW Jr, Dooner HK (1992) Germinal and somatic activity of the maize element Activator (Ac) in Arabidopsis. Genetics 131: 449–459

Kricker MC, Drake JW, Radman M (1992) Duplication-targeted DNA methylation and mutagenesis in the evolution of eukaryotic chromosomes. Proc Natl Acad Sci USA 89: 1075–1079

Lissouba P, Mousseau J, Rizet G, Rossignol J-L (1962) Fine structure of genes in Ascobolus immersus. Adv Genet 11: 343–380

Liu W-M, Schmid CW (1993) Proposed roles for DNA methylation Alu transcriptional and mutational inactivation. Nucleic Acids Res 21: 1351–1359

Martin C, Prescott A, Lister C, MacKay S (1989) Activity of the transposon Tam3 in Antirrhinum and tobacco: possible role of DNA methylation. EMBO J 8: 997–1004

Meyer P, Niedenhof I, ten Lohuis M (1994) Evidence for cytosine methylation of non-symmetrical sequences in transgenic Petunia hybrida. EMBO J 13: 2084–2088

Nelsen-Salz B, Döring H-P (1990) Rare de novo methylation within the transposable element Activator (Ac) in transgenic tobacco plants. Mol Gen Genet 223: 87–96

Nicolas A, Rossignol J-L (1989) Intermediates in homologous recombination revealed by marker effects in Ascobolus. Genome 31: 528–535

Rayssiguier C, Thaler D, Radman M (1989) The barrier to recombination between Escherichia coli and Salmonella typhimurium is disrupted in mismatch-repair mutants. Nature 342: 396–401

Rhounim L, Rossignol J-L, Faugeron G (1992) Epimutation of repeated genes in Ascobolus immersus. EMBO J 11: 4451–4457

Rhounim L, Grégoire A, Salama S, Faugeron G (1994) Clustering of multiple transgene integrations in highly-unstable Ascobolus immersus transformants. Curr Genet 26: 344–351.

Rossignol J-L, Faugeron G (1994) Gene Inactivation triggered by recognition between DNA repeats. Experientia 50:307–317

Rossignol J-L, Nicolas A, Hamza H, Kalogeropoulos A (1988) Recombination and gene conversion in Ascobolus. In: Low B (ed) The Recombination of Genetic Material. Academic, New York, pp 23–72

Schmid C (1991) Human Alu subfamilies and their methylation revealed by blot hybridization. Nucleic Acids Res 19: 5613–5617

Selker EU (1990a) Premeiotic instability of repeated sequences in Neurospora crassa. Annu Rev Genet 24: 579–613

Selker EU (1990b) DNA methylation and chromatin structure: a view from below. Trends Biochem Sci 15: 103–107

Selker EU, Fritz DY, Singer MJ (1993) Dense nonsymmetrical DNA methylation resulting from repeat-induced point mutation in Neurospora. Science 262: 1724–1728

Waldman AS, Liskay RM (1987) Differential effects of base-pair mismatch on intrachromosomal versus extrachromosomal recombination in mouse cells. Proc Natl Acad Sci USA 84: 5340–5344

Wallis JW, Chrebet G, Brodsky G, Rolfe M, Rothstein R (1989) A hyper-recombination mutation in Saccharomyces cerevisiae identifies a novel eukaryotic topoisomerase. Cell 58: 409–419

Gene Silencing in *Drosophila*

S. Henikoff

1 Introduction

"After such a survey one could readily come to the conclusion that the basis of the phenomenon is much too complex for profitable continuation of its investigation at this time". [W.K. Baker (1968), in a review entitled "Position-effect variegation."]

Gene silencing phenomena are now known to be widespread in eukaryotes. But for many prior decades, investigation of gene silencing was largely limited to a phenomenon known as position-effect variegation (PEV) in *Drosophila*. Numerous studies of PEV have led to a wealth of observations resulting in a general set of rules that might lead one to expect that the phenomenon should be well understood (Lewis 1950; Baker 1968; Spofford 1976; Henikoff 1990). However, Baker's lament reveals a frustration that these observations and rules had not elucidated the causal basis for PEV. Nevertheless, progress in understanding genetic mechanisms since Baker's time, in addition to recent insights into PEV and other gene silencing phenomena in flies, have led to a new optimism among

Howard Hughes Medical Institute, Fred Hutchinson Cancer Research Center, Seattle, WA 98104, USA workers in the field. What is also emerging from recent work is that gene silencing phenomena in other organisms might have a similar mechanistic basis to those observed in *Drosophila*.

2 Chromosomal Silencing or Specific Downregulation?

Silencing can occur when a gene is moved to an unaccustomed position, a type of position effect. Position effects are germinally stable, so that many lines appear unchanged even after maintenance in the laboratory for several decades without selection. In *Drosophila*, a commonly observed position effect is up- or downregulation of a transgene that depends on the site of insertion (SPRADLING and RUBIN 1983; LAURIE-AHLBERG and STAM 1987). In these cases, it is thought that nearby regulatory elements affect expression, perhaps involving interactions between the promoter of the inserted gene and transcription factors binding to flanking regulatory sites. These considerations have led to an "enhancer trap" strategy based on position effect that is widely used in studies of developmental regulation (BIER et al. 1989). To the extent that these position effects are caused by elements that regulate specific genes, they might be distinguished from "chromosomal" position effects that occur because a gene is in an unaccustomed chromatin environment. Just what is meant by "environment" is speculative, and it might be that there is no rigid distinction between specific elements that downregulate genes and determinants of chromatin environment. The present discussion of silencing will be limited to chromosomal position-effect phenomena that are thought to reflect environmental differences, with the caveat that until there is a better mechanistic understanding of such differences, this distinction might be artificial. As an operational criterion, chromosomal silencing phenomena usually involve mosaic and random, rather than uniform or patterned, effects on gene expression.

This review will consider several distinguishable position-effect phenomena involving gene silencing. PEV refers specifically to silencing involving incompatibility between euchromatin and heterochromatin, which are contrasting states observed cytologically. Euchromatin, where most of the genes reside, decondenses at interphase and appears less dense at metaphase than heterochromatin. Heterochromatin, where few genes are found, is visible as condensed chromocenters during interphase, typically associated with the nuclear envelope (HESLOP-HARRISON et al. 1993). Heterochromatin in *Drosophila* forms almost entirely at blocks of repetitive DNA, mostly simple sequences of various five to ten base repeats, as well as moderately repetitive sequences apparently derived from transposable elements (JOHN and MIKLOS 1988; LOHE et al. 1993). About the time that heterochromatin was first described (HEITZ 1929), Muller reported PEV as "eversporting displacements" involving chromosome rearrangements that place euchromatic genes adjacent to pericentric heterochromatin (MULLER 1930). It has been thought that the altered chromatin structure characteristics of heterochromatin "spreads" from the rearrangement breakpoint into adjacent euchromatic genes, causing their inactivation (Spofford 1976). PEV also can occur when a gene normally found in heterochromatin is juxtaposed to euchromatin. A different type of position-effect silencing is seen when a transgene inserts onto the tip of a chromosome (Hazelrigg et al. 1984). In each of these examples of silencing, a gene is placed in a position that is known to be different based on cytological criteria. However, for other examples of silencing considered here, there are no obvious cytological differences between the normal and altered environments. Rather, it appears that specific regulatory elements carried on transgenes can mediate silencing interactions between homologous sequences.

3 Position Effects and *trans*-Sensing Effects

In general, a position effect is a change in gene expression based on its altered position in a chromosome. It is possible to prove that a change in expression is a position effect either by removing the sequences causing the position effect or by removing the gene to a new position with restoration of normal function. There are several examples of PEV and telomere position effect in which one or the other of these proofs has been accomplished. For example, PEV has been proven in several cases by obtaining a second rearrangement that separates the affected gene from heterochromatin or by recombination that restores the gene to its normal position on an unrearranged chromosome, accompanied by restoration of full gene activity (Spofford 1976). For genes carried on transposons subject to PEV or telomere position effect, it is generally easy to prove a position effect by transposition to a new site leading to restoration of activity (Levis et al. 1985). It is more frequent, however, that a position effect is assumed without direct proof. Rather, a mutation follows the "rules," so that PEV is inferred. For example, Lindsley and co-workers (1960) found that about 25% of x-ray induced lethal mutations on the X chromosome were suppressible by addition of an extra Y chromosome, which is known to be a general suppressor of PEV (see below). Cytological examination revealed that in every case examined, there was a chromosomal rearrangement placing a site on the X chromosome adjacent to heterochromatin. So although position effect was not proven, nor was it possible to observe mosaicism, the Y-suppressible lethals conformed with rules for PEV, leading the authors to conclude that PEV on vital loci is a frequent result of x-ray induced rearrangement.

Another class of position-dependent phenomena in *Drosophila* result from chromosome pairing interactions *in trans*, typically between allelic genes on homologs, and not from interactions *in cis* with neighboring sequences. Such "*trans*-sensing" effects occur because a gene senses the state of its homolog via somatic pairing, whereby homologous chromosomes are closely associated along their lengths (Tartof and Henikoff 1991). Suspected *trans*-sensing effects are proven by disrupting the association of homologs with concomitant restoration of normal expression, analogous to proofs of position effects *in cis*. While most *trans*-sensing effects are thought to reflect specific gene regulatory mechanisms, others might involve more general chromosomal phenomena. The best understood *trans*-sensing effect is the zeste-white interaction, in which paired wild-type copies of the *white* gene are reduced in expression in the presence of the *zeste*^{1} mutation (Gans 1953). For many years it was assumed that the *zeste* gene is involved in some way in somatic pairing or in gene silencing in general. However, demonstration that zeste protein acts as a positive transcription factor with unusual properties provided an alternative explanation for repression across homologs (Bickel and Pirrotta 1990). *Zeste* protein binds specifically to multiple elements upstream of *white* and other genes, forming multimeric complexes. The *zeste*^{1} mutation increases the size of multimers to a point that contact is made between paired, but not unpaired homologous copies of *white*.

Elucidation of the mechanism underlying the zeste-white interaction provides an important lesson that might be widely applicable to gene silencing phenomena. Silencing per se might not be biologically relevant, but rather a consequence of interfering with a system that has evolved to do something quite different, in this case, activation of specific genes. It must be kept in mind that the study of gene silencing phenomena involves disturbing a gene to the point that a phenotype is seen, so there is no guarantee that silencing is a normal feature of the system in question. This caution applies to PEV and telomere position effect as well. There is no evidence that genes are normally silenced in *Drosophila* because of their position in the genome, such as is the case for X inactivation in placental mammals. As has been shown for *zeste*^{1}, position effect phenomena can provide insight into features of genes and chromosomes that are not readily accessible in other ways, but because these effects are aberrant, the biological relevance of the phenomena cannot be assumed.

4 Using PEV to Study Components of Heterochromatin

Position effects can provide powerful genetic tools for understanding essential features of chromosomes. For example, the study of telomere position effect in yeast has led to tantalizing insights into the role of proteins participating in telomere function, such as RAP1 (Kyrion et al. 1993). Similarly, the study of PEV in *Drosophila* has led to the isolation of genes that encode *trans*-acting factors very likely participating in heterochromatin formation. As this is the most active area of PEV research, involving work in several laboratories, it has been reviewed extensively in recent years (Eissenberg 1989; Grigliatti 1991; Reuter and Spierer 1992). For our purposes, it will suffice to summarize what is known about *trans*-acting factors, using the well-studied HP1 protein as a paradigm for proteins belonging to this class.

HP1 was discovered as a nonhistone chromosomal protein present at high levels in early embryos (JAMES and ELGIN 1986). The protein is present in the single fused chromocenter in polytene salivary nuclei and in the heterochromatin of metaphase chromosomes (POWERS and EISSENBERG 1993). The gene encoding HP1 corresponds to a dosage-dependent suppressor-of-variegation, *Su(var)205* (EISSENBERG et al. 1992). Like other *Su(var)* loci, loss-of-function mutations in *Su(var)205* act as general suppressors of PEV when heterozygous, such that nearly every tested example of PEV is at least partially suppressed in the presence of half the normal dosage of *Su(var)205* (GRIGLIATTI 1991). Conversely, increasing the dosage of the *Su(var)205* gene causes enhancement of PEV (EISSENBERG et al. 1992). Not all *Su(var)* loci show enhancement with increased dosage, suggesting a difference between the fully dosage-sensitive loci such as *Su(var)205*, which might encode specific structural components of heterochromatin, and other *Su(var)*s, which might have other roles (GRIGLIATTI 1991).

Multiple *Su(var)* mutations act additively when in combination (HENIKOFF 1979; GRIGLIATTI 1991). A model accounting for this additivity, as well as for the full dosage sensitivity of *Su(var)*s, is that proteins encoded by these genes form a complex, and it is the concentration of the complex that determines how much heterochromatin can form (LOCKE et al. 1988). If the concentration of the complex is determined by the product of concentrations of each of its components by the mass action law, then a reduction in the dosage of a single component will reduce the concentration of the complex, whereas an increase in dosage will increase complex concentration. Similar considerations apply to fully dosage-sensitive *Enhancer-of-variegation* loci, except that these would participate in complexes that favor euchromatin formation. The success of the mass action model in accounting for additive effects of multiple dosage-dependent loci has led to its wide, though not universal (SPRADLING and KARPEN 1990), acceptance.

Position-effect variegation can also be modified by changing the dosage of the Y chromosome (SPOFFORD 1976), a general feature of the phenomenon known since the early 1930s, when position effect was one of the most widely discussed topics in genetics. While some evidence suggests that the Y chromosome includes specific dosage-sensitive loci encoding *trans*-acting factors (involved in euchromatin formation by the mass action model), other evidence argues that the Y acts as a sink for heterochromatic complexes (DIMITRI and PISANO 1989). This latter possibility is consistent with the mass action model since the Y is a fully heterochromatic chromosome, accounting for more than 20% of all heterochromatin in males, and would be expected to be effective in titrating out complexes needed for heterochromatin formation elsewhere.

So a straightforward picture is emerging from studies of *Su(var)* genes, that the fully dosage-sensitive loci encode subunits of a complex that participates in heterochromatin formation. The question then arises as to how many proteins are present in such a complex. In the absence of biochemical isolation, we can only speculate based on counting *Su(var)*s. In the past, estimates have varied from about 20 to 200. One reason for this wide range is that estimates at the lower end are based on the expected frequency of fully dosage sensitive loci (LOCKE et al. 1988), whereas estimates at the upper end are based on all detected *Su(var)*s regardless of dosage sensitivity (Henikoff 1979; Wustmann et al. 1989). But these estimates assume that screens for *Su(var)*s are sufficiently thorough and sensitive to detect most loci encoding components of the predicted heterochromatic complex. This is an open question, given that genetic screens generally detect only very strong suppressor loci, even though examples of weak suppressors are known (Henikoff 1979). Consequently, all that can be said is that there are probably dozens of different proteins present in the predicted heterochromatic complex, of which only a few are currently known. Of the four *Su(var)* genes whose sequences are currently available, none show protein sequence similarity to one another. However, HP1 has an intriguing similarity to the Polycomb protein, a likely component of a complex involved in developmental silencing of homeotic genes. The alignment comprises a 37 amino acid motif with 24 identities (Paro and Hogness 1991), which suggests that these proteins have similar roles as subunits of their respective complexes. While the cloning of genes based on *Su(var)* phenotype has been illuminating so far, further progress might require advances in vitro methodology. Even then, genetics will be crucial to distinguish between real components of the phenomenon and artifacts.

5 Active Genes in Heterochromatin

It is often assumed that heterochromatin is inert, based on its frequent involvement in gene silencing phenomena. This is untrue. For example, the vast majority of transcription in *Drosophila* comes from the genes for the large ribosomal RNAs, which are located within the pericentric heterochromatin of the X chromosome and on the fully heterochromatic Y. One might object that the nucleolar organizer regions do not qualify as heterochromatic regions because the nucleoli are specialized organelles. Similar objections might be raised concerning several transcribed loci on the Y chromosome shown to specify male fertility factors (Hennig 1990). However, cytogenetic analysis indicates that these are extremely large genes that extend well into fully condensed regions (Gatti and Pimpinelli 1992). In addition, even heterochromatic regions that are not distinctive in any cytological or functional sense have been shown to contain otherwise ordinary genes. A few of these genes have been studied in detail. The *light* gene, located in the distal pericentric heterochromatin of chromosome arm 2L, causes lethality when null and leads to a reduction in eye and malpighian tubule pigmentation when expressed at reduced levels (Lindsley and Zimm 1992). The gene appears to be pleiotropic and is expressed throughout development. Similarly, the *rolled* gene, located in the proximal pericentric heterochromatin of chromosome arm 2R, encodes the *Drosophila* homolog of a tyrosine protein kinase, a regulatory protein that appears to be broadly expressed in flies (Biggs and Zipursky 1992). Several other vital loci have been mapped to heterochromatin, in addition to *concertina*, a maternal effect gene required for egg viability (Lindsley and Zimm

1992). So heterochromatin is not inherently silent. It seems relatively inert because of the low gene density.

Molecular analysis of *light* shows this heterochromatic gene is embedded in moderately repetitive DNA, with numerous such repeats in introns, in contrast to euchromatic genes, which reside in regions with few repetitive sequences (DEVLIN et al. 1990). It would seem that a distinguishing feature of heterochromatic genes is that they have adapted to an environment consisting of high concentrations of repetitive sequences. This apparent adaptation is reflected in the PEV of heterochromatic genes, induced by rearrangements that juxtapose them to distal euchromatic regions. A remarkable inverse response to PEV modifiers is seen: extra Y chromosomes and some *Su(var)*s enhance PEV of heterochromatic genes, whereas lack of the Y in males suppresses, just the opposite of what is consistently seen for euchromatic genes (HEARN et al. 1991). A similar situation appears to apply to centromeres, which invariably are embedded in heterochromatin. In one unique chromosome, a very proximal break that juxtaposed a centromere to euchromatin caused nondisjunction with frequent premature separation of sister chromatids (WINES and HENIKOFF 1992). The effect was enhanced by *Su(var)*s and an extra Y chromosome and suppressed in males lacking a Y, suggesting PEV of a centromere. Thus a variety of functions appear to require a heterochromatic environment, and these are disrupted when moved distally to euchromatic regions.

An important insight that came from studies of heterochromatic gene PEV is that very long-range rather than more local interactions are involved in silencing of these genes (BAKER 1968). For PEV of euchromatic genes, breaks are usually nearby, rarely more than a few hundred kilobases away from the affected gene. In contrast, any proximal break appears to be capable of causing PEV of heterochromatic genes, so long as the gene is moved to a distal position (WAKIMOTO and HEARN 1990). What is important is not whether a heterochromatic gene is moved distally, but whether the heterochromatic block in which it lies is moved far away from other heterochromatic blocks (EBERL et al. 1993). These findings suggest that the cytologically observed chromocenter, comprising heterochromatin joined together from different regions, provides the necessary heterochromatic environment for normal function of heterochromatic genes.

While long-range effects might be interpreted as a feature of PEV specific for heterochromatic genes, recent results have shown that long-range effects also play a role in euchromatic PEV (TALBERT et al. 1994). In this study, a large heterochromatic insertion causing PEV of the euchromatic *brown* gene was found to show weaker PEV when moved distally and stronger PEV when moved proximally as a result of chromosome breaks that were very distant from the insertion itself. Thus the ability of a block of heterochromatin to inactivate a euchromatic gene depends on how close it is to the bulk of heterochromatin in the cell, consistent with the notion that euchromatic genes are inactivated when they are drawn into the chromocenter. PEV might then be seen a nuclear mislocalization phenomenon, with affected heterochromatic genes mislocalized away from the chromocenter (WAKIMOTO and HEARN 1990; EBERL et al. 1993) and affected euchromatic genes targeted into the chromocenter (TALBERT et al. 1994). In both cases gene dysfunction results.

6 "Spreading" Reconsidered

The above hypothesis, that PEV of euchromatic genes such as *white* and *brown* results from their mislocalization of "looping" into the chromocenter, seems at odds with dogma in the field that attributes PEV to the spreading of inactivation along the chromosome (SPOFFORD 1976). Spreading has been proposed to involve initiator elements within heterochromatin and terminator elements within euchromatin (TARTOF et al. 1984; EISSENBERG 1989 ; GRIGLIATTI 1991), even though no such elements have been identified. The evidence for spreading derives in large part from the observation that there is a polarity in the effect of a heterochromatic breakpoint on nearby euchromatic genes, such that a gene closer to heterochromatin is more severely affected in expression that a gene farther away (SPOFFORD 1976). However, the polarity of spreading is also consistent with looping of nearby genes into the condensed chromocenter, where the probability of getting drawn in generally decreases with distance. Spreading and looping are not necessarily incompatible models. For example, X chromosome inactivation in placental mammals might involve long-range looping of special sequences into the X inactivation center followed by the "reeling" in of adjacent regions (RIGGS 1990).

The spreading model has been challenged on other grounds as well. PEV appears to be all-or-none, with completely mutant and completely wild-type spots and clonal patches appearing at random in the affected tissue. It would seem that gene loss could explain such phenotypes, and indeed there is a remarkable resemblance of PEV to transposase-induced gene loss (HENIKOFF et al. 1993). This resemblance has spawned alternatives to the spreading model over the years (SCHULTZ 1936; PROCUNIER and TARTOF 1978; SPRADLING et al. 1993). The most recent is based on a chromosome elimination model for under-representation of heterochromatic sequences in polytene nuclei (SPRADLING et al. 1993). It must be kept in mind that even if elimination is shown to occur in polytene cells, it does not necessarily follow that elimination is the basis for PEV. In fact, there are cases of PEV in which gene loss has been excluded as a mechanism for PEV (HENIKOFF 1981; RUSHLOW et al. 1984; HAYASHI et al. 1990; UMBETOVA et al. 1991). For example, a PEV allele affecting the 2B polytene chromosome puff, seen as compaction of the puff site and failure to puff, showed no reduction in gene copy number when DNA was purified from the tissue in which inactivation was observed (UMBETOVA et al. 1991). Here PEV occurred in polytene cells without gene loss. For genes such as *white* and *brown*, where PEV occurs in tissue that is likely to be diploid, there is even less reason to favor a gene loss mechanism, since underrepresentation of heterochromatin appears not to occur in diploid tissue (SPRADLING et al. 1993). So even if an example of gene loss is found to accompany PEV, the elimination model cannot be a general explanation of the phenomenon.

7 A Relationship Between PEV and Somatic Pairing?

Position-effect variegation models for spreading and mislocalization into (or away from) the chromocenter leave open the question as to what is responsible for the cytological differences between heterochromatin and euchromatin. This question is especially pertinent given that euchromatic regions subject to PEV sometimes appear heterochromatic in polytene nuclei. The normal banding pattern is disrupted, leading to a pebbly appearance typical of distal heterochromatin (HARTMANN-GOLDSTEIN 1966; HENIKOFF 1990). The mechanistic basis for banding in polytene chromosomes is not understood, except that the precise alignment of chromatids and homologous chromosomes plays an important role: all of the strands from both homologs must be precisely paired throughout their length in order to see banding. This correlation between euchromatic and perfect banding, and heterochromatin and disrupted banding led to the idea, now 50 years old, that somatic pairing is a component of PEV (EPHRUSSI and SUTTON 1944). Strong evidence now exists that pairing between homologous chromosomes is responsible for the unusual dominance of *brown* PEV (HENIKOFF et al. 1993). Here it appears that heterochromatin formation at the copy of *brown in cis* to a PEV breakpoint can interact directly with the *trans* copy brought near by somatic pairing of homologs, resulting in *trans*-inactivation of the somatically paired allele. This *trans*-sensing effect is thought to occur because of an unusual feature of a *brown*-specific transcription factor which makes it sensitive to contact with heterochromatin.

Other recent work has implicated somatic pairing of contiguous sequences in the formation of heterochromatin itself. Since heterochromatic sequences consist of large blocks of repetitive DNA, repeated sequences within these blocks might be expected to pair out of register (WAKIMOTO and HEARN 1990). Such associations would result in the disorganized appearance and fusion of the polytene nucleus chromocenter and could be essential for creating a suitable environment for heterochromatic genes and centromeres. Support for this view comes the study of repeat expansions in euchromatin (DORER and HENIKOFF 1994). PEV of the *white* gene occurs at sites of *white*-bearing P-transposons when the transposon is amplified in tandem at the site. As few as three copies of the transposon were associated with variegation that was sensitive to modifiers of PEV, with more extreme variegation occurring with increased copy number. The characteristic PEV phenotypes and the full responses to PEV modifiers indicated that heterochromatin was forming at affected sites. Since the transposon did not include any sequences normally found in heterochromatin, it would seem that heterochromatin formation does not require special sequences. Rather, the self-association of closely linked repeats is sufficient to target the region to the chromocenter, where it becomes part of heterochromatin.

These observations might be relevant to gene silencing phenomena in general. While effects similar to PEV have been seen in other organisms, it has not been possible to infer that heterochromatin was responsible in the absence of previous experience with heterochromatin-induced phenomena or known modifier loci specific for heterochromatin formation. But constitutive heterochromatin is found in all higher eukaryotes; it invariably consists of repetitive DNA and shows very similar cytological characteristics. Thus, parallel observations of gene silencing associated with repeated sequences suggest a similar underlying mechanism for heterochromatin formation. In *Arabidopsis*, silencing was reported to be associated with multiple tandem duplications of a repeated endogenous gene (ASSAAD et al. 1993). This "repeat-induced gene silencing" (RIGS) can likewise be explained as local formation of heterochromatin. Other examples of RIGS in plants might fall into the same category, including transgene repeats causing *trans*-inactivation and co-suppression (MATZKE et al. 1994). Here the multiple copies participating in heterochromatin formation are not closely linked *in cis* to one another, but *in trans* at nonallelic positions. Such participation of *trans* copies in heterochromatin formation has been inferred from studies of *brown trans*-inactivation, in which closely linked duplications of the gene both *in cis* and *in trans* lead to stronger mutant effects (J. Sabl and S.H., unpublished results).

In vertebrates, numerous claims of position effects associated with silencing have been reported in which tandem arrays of transgenes at different sites show a lack of copy number dependence on expression (PALMITER and BRINSTER 1986; Stief et al. 1989). In no case is proof presented that neighboring sequences are involved, leaving open the possibility that silencing is an inherent property of the transgene arrays themselves (DAVIS and MACDONALD 1988). These might be examples of local heterochromatin formation resulting from somatic pairing of the tandem copies and subsequent targeting into chromocenters (DORER and HENIKOFF 1994). Variation in the degree of expression from line to line can then be accounted for by the distance of an insertion site from nearby blocks of heterochromatin, not from interference by neighboring sequences. This is a testable hypothesis, since it predicts a correlation between the degree of silencing and the distance from large heterochromatic blocks for insertions of transgene arrays with the same copy number.

As a corollary to the heterochromatin hypothesis for transgene silencing in vertebrates, sequences that prevent silencing would be those that prevent targeting of the array into a heterochromatic environment. For the chicken lysozyme gene, a matrix or scaffold attachment region (MAR or SAR) was shown to be effective, leading to full copy number dependence of expression when each of the tandem copies included the chicken lysozyme MAR element (STIEF et al. 1989). The interpretation of these experiments was that the MARs were acting as insulators by preventing position effects caused by neighboring sequences. But in *Drosophila*, proven position effects are prevented by flanking insulator elements that have no detectable MAR activity, and conversely, MARs fail to insulate (KELLUM and SCHEDL 1992; CHUNG et al. 1993). Therefore, we have proposed that MARs prevent mislocalization into the chromocenter by attachment of each copy in a tandem array to a fixed matrix, not by preventing interference from neighboring regulatory elements as supposed (DORER and HENIKOFF 1994). Endogenous MARs might similarly prevent heterochromatin formation at natural gene repeats. For example, MARs have been mapped to the *Drosophila* histone gene repeats, present in an array very close to heterochromatin (MIRKOVITCH et al. 1984).

If heterochromatin formation at repeat arrays in euchromatin occurs in vertebrates, it is interesting to consider what might be the fate of trinucleotide repeat expansions that are responsible for several human diseases. Just as higher copy number of a transgene above a threshold is thought to cause the array to be targeted into a chromocenter, similar mislocalization might occur when any sequence becomes locally repetitive. Important factors would be the length of the repeat (or the number of units), the distance from the nearest heterochromatic block along the chromosome, and the distance to flanking MARs that would function to prevent mislocalization. By this model, trinucleotide repeat arrays may be thought of as targeting elements that can lead to heterochromatin formation at sequences mislocalized to a chromocenter. Earlier, LAIRD (1987) had proposed that the fragile-X syndrome, now known to be caused by trinucleotide repeat expansion, involves heterochromatin formation. The basis of this disease could be mislocalization of the fragile-X region into a chromocenter.

8 Telomere Position Effect

Compared to the extensive literature concerning PEV, there are only a few published studies describing telomere position effect in *Drosophila*. One report demonstrated that when a *white* gene carried on a P-transposon was mobilized to a subtelomeric region of chromosome arm 3R, it showed variegated expression that depended on its position near the telomere (HAZELRIGG et al. 1984). More recent mapping of the site of insertion shows that it lies about 10 kb from the most distal transcription unit on 3R and about 16 kb from the tip, embedded within an array of repeats (LEVIS et al. 1993). Another example involved a P-transposon bearing the *rosy* gene, which showed sharply reduced expression when inserted into a related set of repeats at the end of the X chromosome (KARPEN and SPRADLING 1992). An apparent inconsistency between the two reported cases is that the *white* telomeric transgene is completely insensitive to PEV modifiers (TALBERT et al. 1994), whereas the *rosy* transgene is Y-suppressible (KARPEN and SPRADLING 1992). A possible explanation for this difference is that the *rosy* transgene was situated on a mini-chromosome, and was therefore only about 250 kb away from a one megabase block of pericentric heterochromatin. By the nuclear localization model (TALBERT et al. 1994), the flanking repeats would form paired structures which target the region to the chromocenter, so that *rosy* "telomere position effect" would actually be an example of PEV.

The insensitivity of *white* telomeric position effect to PEV modifiers indicates that there are different protein components necessary for telomere formation than for heterochromatin formation. This difference might be dictated by the need to keep the telomeres in their accustomed place at the other end of the nucleus from the heterochromatin (Rabl 1885; Hochstrasser et al. 1986). Telomeric repeats are not essential in *Drosophila*, since terminal deletions that remove them do not interfere with chromosome function (Levis 1989; Biessmann et al. 1990). The repeats do appear to be necessary for telomere position effect, since removal of repeats is correlated with phenotypic suppression (Levis 1989; Levis et al. 1993). Silencing might result from nuclear mislocalization, as hypothesized for PEV. It is worth noting that genes placed adjacent to telomeres in yeast are variably repressed. This telomere position effect might similarly require the $C_{1-3}A$ repeats at chromosome ends (J. Stavenhagen and V. Zakian, personal communication), although a direct test of this possibility is difficult because removal of the repeats from linear chromosomes leads to chromosome loss (Sandell and Zakian 1993).

9 Transgene Silencing Mediated by Specific Regulatory Elements

As mentioned at the outset, it may be difficult to distinguish between chromosomal silencing and gene-specific downregulation, except in cases involving cytologically observable elements such as heterochromatin and telomeres. However, the discovery that known regulatory elements can mediate silencing suggests that this distinction is somewhat artificial. An example is the zeste-white interaction, in which paired copies of *white* are repressed because of a mutant transcriptional activator protein. More recently, it was shown that homologous pairing-dependent repression of a *white* gene lacking upstream *cis*-regulatory elements (mini-*white*) occurred when linked to a 2-kb regulatory element derived from the *engrailed* (*en*) locus (Kassis et al. 1991). Paired, but not unpaired, copies of *en*-mini-*white* caused repression of both copies in a *zeste*-independent fashion. This was not strictly an allelic interaction, since repression also occurred for copies inserted at nearby sites (on a cytological scale) but not for copies at distant or unlinked sites. This remarkable feature of the *en* regulatory region is not unique. Similar *trans*-sensing repression has been reported for a regulatory element present in the *polyhomeotic (ph)* gene, a member of the *Polycomb (Pc)* group (Fauvarque and Dura 1993).

Somatic pairing across homologs mediated by regulatory elements might be analogous to the somatic pairing proposed to underlie heterochromatin formation. The *ph* gene is a member of the *Pc* group of genes needed for the maintenance of the repressed state of homeotic genes through development (Paro 1990). Silencing mediated by the *ph* element depends upon the activity of genes in the *Pc* group, including *ph* itself, but generally not *Su(var)*s, indicating that different components are involved in *Pc*-mediated repression and heterochromatin formation. Nevertheless, there are striking parallels between the two genetic systems. Like the *Su(var)* genes, the *Pc* group genes appear to behave in accordance with the mass-action model, as if complex formation underlies action (DeCamillis et al. 1992). In support of this model, Polycomb and polyhomeotic proteins co-localize at specific sites in polytene chromosomes (DeCamillis et al. 1992), causing silencing and chromatin structural alterations over a region on the order of hundreds of kilobases (Orlando and Paro 1993). As for PEV, in which a class of enhancer loci have been identified which act oppositely so *Su(var)*s in a dosage-dependent manner, the *trithorax* group genes act oppositely to genes in the *Pc* group (Paro 1990). In this way, two very different systems with different components might act similarly in causing silencing.

The transgene examples illustrate that the *en* and *ph* regulatory elements play a role in interactions between homologs, but it is unclear just what that role is. One possibility is that the proteins mediating the interaction form larger complexes that interfere with transgene expression similar to what has been shown for the zeste protein (Kassis 1994). Another possibility is that pairing itself leads to transcriptional interference. Along these lines, it is interesting that the repressed phenotype seen for *en*-mini-*white* is mimicked by mini-*white* alone when duplicated *in cis* in a head-to-head fashion (Dorer and Henikoff 1994). Duplication-based repression is not dependent on position, since it is seen for the same duplicated element at several different sites. Head-to-tail copies of mini-*white* show much weaker repression. The similarity of the repressed phenotype of *en*-mini-*white*, and the dependence on repeat orientation, suggests that somatic pairing is occurring between copies. A more severe effect might occur for a hairpin, which would pair along its full length, than for the distorted loop, which would form for a tandem duplication (Gubb et al. 1990). In this case there is no known regulatory element present that could mediate repression. Thus, it would appear that simply bringing copies of mini-*white* together can cause interference in expression.

10 Conclusion

Gene silencing in *Drosophila* encompasses several different phenomena, including PEV, telomere position effect, and repression *in trans* mediated by regulatory elements. While all of these silencing effects are aberrant, and therefore might not cause silencing in normal flies, all provide important clues to the nature of interactions between chromosomes and the proteins and genetic elements that might mediate these interactions. Study of these effects in *Drosophila*, with its powerful genetics and unsurpassed ability to manipulate genes and chromosomes, is providing clues to chromosome and nuclear function and organization that have not been possible in studies of other higher eukaryotes. It seems likely that progress in *Drosophila* will help to elucidate gene silencing phenomena in general.

Acknowledgements. Ideas expressed here drew heavily on discussions with my colleagues Doug Dorer, Bob Levis, Joy Sabl, Paul Talbert, and Barbara Wakimoto as well as various of their experimental observations. Support was provided by the Howard Hughes Medical Institute.

References

Assaad FF, Tucker KL, Signer ER (1993) Epigenetic repeat-induced gene silencing (RIGS) in Arabidopsis. Plant Mol Biol 22: 1067–1085

Baker WK (1968) Position-effect variegation. Adv Genet 14: 133–169

Bickel S, Pirrotta V (1990) Self-association of the Drosophila zeste protein is responsible for transvection effects. EMBO J 9: 2959–2967

Bier E, Vaessin H, Shepherd S, Lee K, McCall K, Barbel S, Ackerman L, Carretto R, Uemura T, Grell E, Jan LY, Jan YN (1989) Searching for pattern and mutation in the Drosophila genome with a P-lacZ vector. Genes Dev 3: 1273–1287

Biessmann H, Carter SB, Mason JM (1990) Chromosome ends in Drosophila without telomeric DNA sequences. Proc Natl Acad Sci USA 87: 1758–1761

Biggs WH III, Zipursky SL (1992) Primary structure, expression and signal-dependent tyrosine phosphorylation of a Drosophila homolog of extracellular signal-regulated kinase. Proc Natl Acad Sci USA 89: 6295–6299

Chung JH, Whiteley M, Felsenfeld G (1993) A 5′ element of the chicken beta-globin domain serves as an insulator in human erythroid cells and protects against position effect in Drosophila. Cell 74: 505–514

Davis BP, MacDonald RJ (1988) Limited transcription of rat elastase I transgene repeats in transgenic mice. Genes Dev 2: 13–22

DeCamillis M, Cheng N, Pierre D, Brock HW (1992) The polyhomeotic gene of Drosophila encodes a chromatin protein that shares polytene chromosome-binding sites wth Polycomb. Genes Dev 6: 223–232

Devlin RH, Bingham B, Wakimoto B (1990) The organization and expression of the light gene, a heterochromatic gene of Drosophila melanogaster. Genetics 125: 129–140

Dimitri P, Pisano C (1989) Position effect variegation in Drosophila melanogaster: relationship between suppression effect and the amount of Y chromosome. Genetics 122: 793–800

Dorer DR, Henikoff S (1994) Expansions of transgene repeats cause heterochromatin formation and gene silencing in Drosophila Cell 77: 993–1002

Eberl DF, Duyf BJ, Hilliker AJ (1993) The role of heterochromatin in the expression of a heterochromatic gene, the rolled locus of Drosophila melanogaster. Genetics 134: 277–292

Eissenberg JC (1989) Position effect variegation in Drosophila: towards a genetics of chromatin assembly. Bioessays 11: 14–17

Eissenberg JC, Morris GD, Reuter G, Hartnett T (1992) The heterochromatin-associated protein HP-1 is an essential protein in Drosophila with dosage-dependent effects on position-effect variegation. Genetics 131: 345–352

Ephrussi B, Sutton E (1944) A reconsideration of the mechanism of position effect. Proc Natl Acad Sci USA 30: 183–197

Fauvarque M-O, Dura J-M (1993) polyhomeotic regulatory sequences induce developmental regulator-dependent variegation and targeted P-element insertions in Drosophila. Genes Dev 7: 1508–1520

Gans M (1953) Etude genetique et physiologique du mutant z de Drosophila melanogaster. Bull Biol Fr Belg Suppl 38: 1–90

Gatti M, Pimpinelli S (1992) Functional elements in Drosophila melanogaster heterochromatin Annu Rev Genet 26: 239–275

Grigliatti T (1991) Position-effect variegation—an assay for nonhistone chromosomal proteins and chromatin assembly and modifying factors. In: Hamkalo BA, Elgin SCR (eds) Functional organization of the nucleus: a laboratory guide. Academic, London, pp 587–627

Gubb D, Ashburner M, Roote J, Davis T (1990) A novel transvection phenomenon affecting the white gene of Drosophila melanogaster. Genetics 126: 167–176

Hartmann-Goldstein IJ (1966) Relationship of heterochromatin to puffs in a salivary gland chromosome of Drosophila. Naturwissenschaften 53: 91

Hayashi S, Ruddell A, Sinclair D, Grigliatti T (1990) Chromosomal structure is altered by mutations that suppress or enhance position effect variegation. Chromosoma 99: 391–400

Hazelrigg T, Levis R, Rubin GM (1984) Transformation of white locus DNA in Drosophila: dosage compensation, zeste interaction, and position effects. Cell 36: 469–481

Hearn MG, Hedrick A, Grigliatti TA, Wakimoto BT (1991) The effect of modifiers of position-effect variegation on the variegation of heterochromatic genes of Drosophila melanogaster. Genetics 128: 785–797
Heitz E (1929) Heterochromatin, Chromocentren, Chromomeren. Ber Dtsch Bot Ges 47: 274–284
Henikoff S (1979) Position effects and variegation enhancers in an autosomal region of Drosophila melanogaster. Genetics 93: 105–115
Henikoff S (1981) Position-effect variegation and chromosome structure of a heat shock puff in Drosophila. Chromosoma (Berl) 83: 381–393
Henikoff S (1990) Position-effect variegation after 60 years. Trends Genet 6: 422–426
Henikoff S, Loughney K, Dreesen TD (1993) The enigma of dominant position-effect variegation in Drosophila. In: Heslop-Harrison JS (ed) The chromosome. BIOS, Oxford, pp 183–196
Hennig W (1990) The Y chromosome of Drosophila. In: Adolph KW (ed) Chromosomes: eukaryotic, procaryotic, and viral. CRC Press, Boca Raton, pp 213–238
Heslop-Harrison JS, Leitch AR, Schwarzacher T (1993) The physical organization of interphase nuclei. In: Heslop-Harrison JS, Flavell RB (eds) The chromosome. BIOS, pp 221–232
Hochstrasser M, Mathog D, Gruenbaum Y, Saumweber H, Sedat JW (1986) Spatial organization of chromosomes in the salivary gland nuclei of Drosophila melanogaster. J Cell Biol 102: 112–123
James TC, Elgin SCR (1986) Identification of a nonhistone chromosomal protein associated with heterochromatin in Drosophila and its gene. Mol Cell Biol 6: 3862–3872
John B, Miklos GLG (1988) The eukaryote genome in development and evolution. Allen and Unwin, Boston
Karpen GH, Spradling AC (1992) Analysis of subtelomeric heterochromatin in the Drosophila minichromosome Dp1187 by single P element insertional mutagenesis. Genetics 132: 737–753
Kassis JA (1994) Unusual properties of regulatory DNA from the Drosophila engrailed gene: three "pairing-sensitive" sites within a 1.6-kb region. Genetics 136: 1025–1038
Kassis JA, VanSickle EP, Sensabaugh SM (1991) A fragment of engrailed regulatory DNA can mediate transvection of the white gene in Drosophila. Genetics 128: 751–761
Kellum R, Schedl P (1992) A group of scs elements function as domain boundaries in an enhancer-blocking assay. Mol Cell Biol 12: 2424–2431
Kyrion G, Liu K, Liu C, Lustig AJ (1993) RAP1 and telomere structure regulate telomere position effects in Saccharomyces cerevisiae. Genes Dev 7: 1146–1159
Laird CD (1987) Proposed mechanism of inheritance and expression of the human fragile-X syndrome of mental retardation. Genetics 117: 587–599
Laurie-Ahlberg CC, Stam LF (1987) Use of P-element-mediated transformation to identify the molecular basis of naturally occurring variants affecting Adh expression in Drosophila melanogaster. Genetics 115: 129–140
Levis RW (1989) Viable deletions of a telomere from a Drosophila chromosome. Cell 58: 791–801
Levis R, Hazelrigg T, Rubin GM (1985) Effects of genomic position on the expression of transduced copies of the white gene of Drosophila. Science 229: 558–561
Levis RW, Ganesan R, Houtchens K, Tolar LA, Sheen F-m (1993) Transposons in place of telomeric repeats at a Drosophila telomere. Cell 75: 1083–1093
Lewis EB (1950) The phenomenon of position effect. Adv Genet 3: 73–115
Lindsley DL, Zimm GG (1992) The genome of Drosophila melanogaster. Academic, San Diego
Lindsley DL, Edington CW, von Halle ES (1960) Sex-linked recessive lethals in Drosophila whose expression is suppressed by the Y chromosome. Genetics 45: 1649–1670
Locke J, Kotarski MA, Tartof KD (1988) Dosage-dependent modifiers of position effect variegation in Drosophila and a mass action model that explains their effect. Genetics 120: 181–198
Lohe A, Hilliker AJ, Roberts PA (1993) Mapping simple repeated DNA sequences in heterochromatin of Drosophila melanogaster. Genetics 134: 1149–1174
Matzke M, Matzke AJM, Scheid OM (1994) Inactivation of repeated genes—DNA–DNA interaction? In: Paszkowski (ed) Homologous recombination and gene silencing in plants. Kluwer, Amsterdam, pp 271–307
Mirkovitch J, Mirault M-E, Laemmli UK (1984) Organization of the higher-order chromatin loop: specific DNA attachment sites on the nuclear scaffold. Cell 39: 223–232
Muller HJ (1930) Types of visible variations induced by X-rays in Drosophila. J Genet 22: 299–334
Orlando V, Paro R (1993) Mapping Polycomb-repressed domains in the bithorax complex using in vivo formaldehyde cross-linked chromatin. Cell 75: 1187–1198
Palmiter RD, Brinster RL (1986) Germ-line transformation of mice. Annu Rev Genet 20: 465–499
Paro R (1990) Imprinting a determined state into the chromatin of Drosophila. Trends Genet 6: 416–421
Paro R, Hogness DS (1991) The Polycomb protein shares a homologous domain with a heterochromatin-associated protein of Drosophila. Proc Natl Acad Sci USA 88: 263–267

Powers JA, Eissenberg JC (1993) Overlapping domains of the heterochromatin-associated protein HP1 mediate nuclear localization and heterochromatin binding. J Cell Biol 120: 291–299
Procunier JD, Tartof KD (1978) A genetic locus having trans and contiguous cis functions that control the disproportional replication of ribosomal RNA genes in Drosophila melanogaster. Genetics 88: 67–79
Rabl C (1885) Über Zellteilung. Morphol Jahrb 10: 214–230
Reuter G, Spierer P (1992) Position effect variegation and chromatin proteins. Bioessays 14: 605–612
Riggs AD (1990) DNA methylation and late replication probably aid cell memory, and type I DNA reeling could aid chromosome folding and enhancer function. Philos Trans R Soc Lond [Biol] 326: 285–297
Rushlow CA, Bender W, Chovnick A (1984) Studies on the mechanism of heterochromatic position effect at the rosy locus of Drosophila melanogaster. Genetics 108: 603–615
Sandell LL, Zakian VA (1993) Loss of a yeast telomere: arrest, recovery and chromosome loss. Cell 75: 729–739
Schultz J (1936) Variegation in Drosophila and the inert heterochromatic regions. Proc Natl Acad Sci USA 22: 27–33
Spofford JB (1976) Position-effect variegation in Drosophila. In: Ashburner M, Novitski E (eds) Genetics and biology of Drosophila, vol 1c. Academic, San Diego, pp 955–1019
Spradling AC, Karpen GH (1990) Sixty years of mystery. Genetics 126: 779–784
Spradling AC, Rubin GM (1983) The effect of chromosomal position on the expression of the Drosophila xanthine dehydrogenase gene. Cell 34: 47–57
Spradling AC, Karpen G, Glaser R, Zhang P (1993) Evolutionary conservation of developmental mechanisms: DNA elimination in Drosophila. In: Spradling AC (ed) Evolutionary conservation of developmental mechanisms Wiley-Liss, New York, pp 39–53
Stief A, Winter DM, Stratling WH, Sippel AE (1989) A nuclear DNA attachment element mediates elevated and position-independent gene activity. Nature 341: 343–345
Talbert PB, LeCiel CDS, Henikoff S (1994) Modification of the Drosophila heterochromatic mutation brown Dominant by linkage alterations. Genetics 136: 559–571
Tartof KD, Henikoff S (1991) Trans-sensing effects from Drosophila to humans. Cell 65: 201–203
Tartof KD, Hobbs C, Jones M (1984) A structural basis for variegating position effects. Cell 37: 869–878
Umbetova GH, Belyaeva ES, Baricheva EM, Zhimulev IF (1991) Cytogenetic and molecular aspects of position effect variegation in Drosophila melanogaster IV. Underreplication of chromosomal material as a result of gene inactivation. Chromosoma 101: 55–61
Wakimoto B, Hearn M (1990) The effects of chromosome rearrangements on the expression of heterochromatic genes in chromosome 2L of Drosophila melanogaster. Genetics 125: 141–151
Wines DR, Henikoff S (1992) Somatic instability of a Drosophila chromosome. Genetics 131: 683–691
Wustmann G, Szidonya J, Taubert H, Reuter G (1989) The genetics of position-effect variegation modifying loci in Drosophila melanogaster. Mol Gen Genet 217: 520–527

Uptake of Foreign DNA by Mammalian Cells Via the Gastrointestinal Tract in Mice: Methylation of Foreign DNA—A Cellular Defense Mechanism

W. Doerfler

1 Introduction

Many reviews have been published on eukaryotic DNA methylation, but in spite of intensive work for more than two decades, many of the biological implications of this genetic signal, by some termed an "epigenetic signal", still remain to be elucidated. The nucleotide 5-methyldeoxycytidine (5-mC) has been recognized as a modulator of DNA motif–protein interactions and hence has the potential to affect many reactions of the DNA molecule. For my own motivation to continue research in this fascinating yet complex field of investigations, the finding has been decisively stimulating that in distinct, randomly selected segments of the human genome patterns of DNA methylation are identical at the nucleotide level or in larger regions among different individuals of diverse ethnic backgrounds (Kochanek et al. 1990, 1993; Behn-Krappa et al. 1991). Therefore, it is highly unlikely that patterns of DNA methylation in established mammalian genomes are the result of a randomly acting DNA methyltransferase system. The available

Institut für Genetik, Universität zu Köln, Weyertal 121, 50931 Köln, Germany

evidence indicates that patterns of DNA methylation are species-, cell type- and DNA segment-specific and vary with stage of development. We do not yet know how these patterns arise, whether they can fluctuate, or what their functional significance may be. Most researchers in the field agree that motif-specific methylation serves an important function in the long-term silencing of eukaryotic promoters (DOERFLER 1983).

Since meaningful work on DNA methylation is conceptually complex and technically difficult, particularly because of the necessary application of genomic sequencing techniques, many have tried to stay away from studies on the meaning of the unexpected fifth nucleotide in the genetic repertoire of mammalian DNA. Nevertheless, important fields of research such as work with transgenic organisms and recently, research towards effective gene therapy for medical applications have encountered de novo methylation as a mechanism affecting the expression of foreign DNA integrated into established mammalian genomes. This mechanism often thwarts the meaningful functional manipulations of these genomes by shutting off the newly introduced foreign DNA. So far, the realm of research that pertains to changes in methylation patterns in established genomes at sites far remote from the locus of foreign DNA insertion has been almost totally neglected. These poorly understood consequences of foreign DNA insertion could render current interpretations of work on transgenic organisms ambiguous and threaten to invalidate some of the simpler regimes for gene therapy.

For our own work on various aspects of eukaryotic DNA methylation, we have concentrated on virus systems, and in particular on mammalian cells transformed by adenoviruses or on adenovirus type 12 (Ad12)-induced hamster tumor cells and on detailed analyses of patterns of DNA methylation in distinct segments of the human genome. In this review, I will not repeat the summaries of our work of over a decade (between 1978 and 1990) on promoter inactivation by sequence-specific promoter methylation (SUTTER and DOERFLER 1979, 1980; DOERFLER 1981, 1983, 1989, 1993, for reviews). In a recent report, we were able to correlate the inactivating effect of sequence-specific methylation on eukaryotic (viral and mammalian) promoters quantitatively not only with the number of methylatable 5′-CG-3′ sequences in a promoter (MUIZNIEKS and DOERFLER 1994a), but also to changes in DNA bending in the E2A late promoter of adenovirus type 2 (Ad2) DNA (MUIZNIEKS and DOERFLER 1994b).

Three topics will constitute the focus of this overview of our own research:

1. The problem of de novo methylation of foreign DNA integrated into an established mammalian genome can be studied with integrated adenovirus DNA in tumors and in cell lines. Early after the infection of hamster cells with Ad12, large amounts of the viral (foreign) DNA become linked to chromosomal DNA (DOERFLER 1968, 1970; SCHRÖER and DOERFLER 1995). The bulk of our work has been performed with cloned lines of Ad12-induced hamster tumor cells or Ad2 or Ad12-transformed cell lines (SUTTER et al. 1978; VARDIMON et al. 1980; KRUCZEK and DOERFLER 1982; KUHLMANN and DOERFLER 1982; OREND et al. 1991, 1995).

2. As one of the consequences of adenovirus DNA insertion and transformation, we have documented changes in DNA methylation in cellular DNA sequences at the site of adenovirus DNA insertion (LICHTENBERG et al. 1988), and in trans far remote from the chromosomal location of the viral DNA in cellular DNA segments on many different chromosomes.
3. The uptake of foreign DNA has also been investigated in living organisms like the mouse. We have demonstrated (SCHUBBERT et al. 1994; SCHUBBERT and DOERFLER 1995) that unprotected phage M13 DNA is not completely degraded upon passage through the gastrointestinal system of the mouse. Moreover, fragments of about 900 base pairs (bp) can be transiently recovered from the blood of mice fed with M13 DNA, and this DNA has been retrieved in the nuclei and cytoplasm of peripheral white blood cells of these animals. We are currently investigating whether the foreign DNA that has somehow penetrated the gastrointestinal barrier of mice can become integrated into cellular DNA. These results imply that living organisms are incessantly exposed to the challenge of having to deal with foreign DNA which is amply ingested with the daily food supply.

2 De Novo Methylation of Foreign DNA—A Cellular Defense Mechanism

In the course of studies on the integration of Ad12 DNA and its differential expression of early (expressed) versus late (not expressed) viral genes in transformed hamster cell lines or in Ad12-induced hamster tumor cell lines, we discovered an inverse relationship between integrated viral gene expression and the extent of DNA methylation (SUTTER and DOERFLER 1979, 1980; VARDIMON et al. 1980; KRUCZEK and DOERFLER 1982; TOTH et al. 1989). Hence, some of our work on DNA methylation has been performed with the aim of improving our understanding of the mechanism of adenovirus oncogenesis. Many aspects of subsequent studies have gone beyond this interesting biological problem (DOERFLER 1981, 1983, 1984, 1989).

Since foreign DNA inserted into established genomes has frequently been found to become de novo methylated, I have proposed (DOERFLER 1991) that de novo methylation of transgenes and foreign DNA can be viewed as a cellular defense mechanism against the activity of foreign DNA that has intruded into an established genome. This intrusion may occur either as a consequence of natural events and selection or as the result of transfection experiments. The de novo methylation of foreign DNA has been documented in both mammalian and plant systems (for pertinent reviews in plant biology, see Meyer, Fedoroff, and Singer and Selker, this volume). It has also been hypothesized that existing patterns of DNA methylation in the genomes of organisms might—at least in part—represent the sum total of ancient integration and de novo methylation events which have, of course, been modified by numerous selection processes.

Pursuing the idea of a cellular defense mechanism, the question arose as to where the most likely portal of entry of foreign DNA into an organism should be sought, because if would be unrealistic to presume that defense and selection mechanisms would not continue in the organisms of the present time. Obviously, the gastrointestinal system of all organisms is constantly exposed to a large amount of foreign DNA, most of which is complexed with proteins but will be liberated partly as free DNA during food digestion. It was out of these considerations that the experiments on the fate of free DNA in the murine gastrointestinal tract were performed (see Sect. 5). Since foreign DNA can enter the animals' bloodstream, de novo methylation of integrated foreign DNA could in fact constitute an ancient defense mechanism, perhaps comparable to the animals' immune system. The gastrointestinal tract is, of course, also the organ system with maximal exposure to foreign proteins and infectious agents and hence harbors the bulk of the organism's immune system.

3 De Novo DNA Methylation of Integrated Foreign DNA

3.1 General Remarks

In our investigations on the structure of integrated Ad12 or Ad2 genomes in transformed hamster cells, we found early on that these genomes became de novo methylated in very specific patterns (Sutter et al. 1978; Sutter and Doerfler 1979, 1980; Vardimon et al. 1980; Kuhlmann and Doerfler 1982; Orend et al. 1991, 1995). We had previously shown that the virion DNA from purified Ad12 particles was not detectably methylated (Günthert et al. 1976; Wienhues and Doerfler 1985; Kämmer and Doerfler 1995). Thus, it was possible to unequivocally demonstrate that the previously unmethylated Ad12 DNA was de novo methylated after integration into the cellular genome.

An inverse correlation between the extent of DNA methylation and the level of transcription has been observed in many parts of integrated adenovirus genomes (Sutter and Doerfler 1979, 1980). This correlation has subsequently been refined for the promoter regions of integrated adenovirus genes (Kruczek and Doerfler 1982). These observations have initiated a decade of research on the role that sequence-specific promoter methylation can play in the long-term silencing of eukaryotic genes. In these studies we have mainly, but not exclusively, used viral promoters (for reviews, see Doerfler 1981, 1983, 1984, 1989, 1992, 1993; Doerfler et al. 1993). The importance of DNA methylation for long-term gene inactivation has more recently been recognized, also in developmental biology and in human genetics. Earlier work on viral promoters has conceptually opened the path for studies on complex genetic phenomena, e.g., genomic imprinting in mammalian genomes (Surani et al. 1984; Sapienza et al. 1987; Reik et al. 1987; Swain et al. 1987; Li et al. 1993). The observation that integrated foreign

DNA molecules like the Ad12 genome can be de novo methylated and consequently partly or completely inactivated has not been restricted to adenovirus genomes. Other integrated viral genomes, or for that matter any foreign DNA integrated into established genomes, e.g., after transfection and selection in mice (LETTMANN et al. 1991) or in plants (LINN et al. 1990), have very frequently been subjected to the same, apparently ubiquitous control mechanism and have become extensively de novo methylated and inactivated.

The genome of Epstein-Barr virus (EBV), a member of the herpesvirus group, which can persist in virus-transformed cells predominantly in a nonintegrated, circular, episomally free form, can also become de novo methylated in specific patterns (ERNBERG et al. 1989; HU et al. 1991). Thus de novo methylation of foreign genomes in mammalian cells is not solely associated with the integrated state of the newly acquired DNA. Persisting EBV genomes continue to be replicated in synchrony with the cycle of the cellular genome, presumably by the cellular DNA replication systems. It is challenging to consider the possibility that the cellular replication machinery may be intimately associated with the apparatus for the de novo methylation of DNA which would then be responsible for the methylation of the EBV DNA. In contrast, free adenovirus DNA replication is at least partly self-sufficient and provides its own replication system, which, however, presumably lacks DNA methyltransferase activities. Perhaps, for that reason, intracellular, free adenovirus DNA has never been found to become de novo methylated (WIENHUES and DOERFLER 1985; KÄMMER and DOERFLER 1995).

Since the insertion of foreign DNA into established genomes and its continued transcription constitutes a major goal of many, though not all, strategies in gene therapy, the mechanism of de novo methylation and subsequent long-term inactivation of integrated foreign genomes requires serious consideration and detailed investigations. De novo methylation may represent a major obstacle in this frequently considered path towards the successful repair of genetic defects in mammalian cells. Alternative approaches such as the presentation of foreign genes in free nonintegrated form, e.g., in free adenovirus genomes (RAGOT et al. 1993), under conditions in which they do not predominantly integrate, may have a better chance of providing means for the long-term, nonobstructed expression of foreign genes designed to substitute for missing genetic functions in a cell or in an organism. Even if one could replace a defective gene exactly by the wild-type allele, the question must be raised as to whether this replaced gene or DNA segment will also be de novo methylated because it does not carry the authentic cell-specific methylation pattern and will thus be recognized by the de novo methylation system of the cell.

As mentioned earlier, the de novo methylation of integrated foreign DNA in established genomes can be viewed as an ancient cellular defense mechanism which apparently can operate selectively. Possibly by survival and selection of cells with an optimized set of de novo methylated and inactivated or nonmethylated and continuously expressing foreign genes, specific patterns of these foreign, de novo methylated genes persist and contribute to the constellation of newly introduced genes in a thus altered genome. Adenovirus-transformed cells provide an example of this mechanism. Frequently, the early viral genes, mainly the E1 and E4 gene segments, do not succumb to this cellular defense mechanism (OREND et al. 1991, 1995), probably because by selection they can escape inactivation and con-tribute to the transformed state of those cells in which they continue to be expressed.

It appears that cells have developed different defense mechanisms against the onslaught of foreign DNA. Under experimental conditions, a variety of options have become available for the introduction of foreign DNA into cells in culture. It is unknown how frequently the cells of an intact organism are exposed to, take up, and chromosomally integrate foreign DNA. In Sect. 5, an experimental approach towards answering some of these questions will be described. It is likely that the cytoplasmic membrane is a first barrier against the penetration of foreign DNA molecules. Once that barrier has been overcome, foreign DNA can be nucleo-lytically degraded in the cytoplasm or in its organelles. Nevertheless, foreign DNA can be transported to the nucleus and become integrated. Such integrated genomes can be lost again from the cellular genome, as exemplified by the existence of morphological revertants of Ad12-transformed cells in which viral genomes, in part or in toto, have been excised (GRONEBERG et al. 1978; EICK et al. 1980). Finally, in case all these possibilities to eliminate foreign genes had failed, they could eventually become de novo methylated and thus inactivated. We know very little about the mechanism of de novo methylation in mammalian cells and have, therefore, continued to study this process with integrated adenovirus genomes.

3.2 Initiation of De Novo Methylation in Mammalian Cells

Integrated Ad12 DNA in hamster tumor cells or Ad12 DNA fixed in the hamster cell genome by transfection and selection for another marker is not immediately de novo methyl-ated. It requires an unknown number of cell generations—and other unknown factors—to initiate de novo methylation. We have investigated where de novo methylation is initiated in the colinearly integrated Ad12 genome. It commences in two paracentrally located regions of Ad12 DNA and not, for example, at the termini of Ad12 DNA, which are linked to cellular DNA sequences that probably already present an established methylation pattern. Methylation then spreads from the sites of initiation and proceeds to large parts of the integrated viral genomes in the tumor-derived or Ad12-transformed hamster cell lines.

There is evidence from several different systems in which de novo methylation has been studied that certain nucleotide sequences may be preferentially and initially de novo methylated (SZYF et al. 1990; MUMMANENI et al. 1993; HASSE and SCHULZ 1994). While nucleotide sequences will play a certain role in selecting sites for the initiation of de novo methylation, our results from investigations on de novo methylation in the adenovirus system argue that nucleotide sequence by itself cannot be the sole determining parameter that characterizes sites of initiation of de novo methylation:

1. In integrated Ad12 genomes, de novo methylation is initiated in the paracentrally located nucleotide 20 885–24 053 fragment of the viral genome, more precisely in an internal segment of this region (Orend et al. 1995). When the same viral DNA segment is transposed, e.g., to the left end of the integrated Ad12 genomes in cell line T637, the transposed Ad12 DNA segment with the same nucleotide sequence is not methylated or at least not to the same extent as the internally located segment is.
2. When fragments of Ad12 DNA such as the nucleotide 1–5574 EcoRI-C fragment or the nucleotide 20 885–24 053 (PstI-D) fragment are transfected into and fixed by integration in the genomes of mammalian cells, these DNA fragments become methylated in some cell lines, but remain unmethylated in others, possibly dependent on the site of foreign DNA integration (Orend et al. 1995). In contrast, cells that have been transformed by infection with Ad12 virions carry the entire Ad12 genome in an integrated form, and the 1–5574 nucleotide fragment remains hypomethylated, while the internal PstI-D fragment becomes heavily methylated.
3. In the Ad2-transformed hamster cell line HE1 (Cook and Lewis 1979), the late E2A promoter in the Ad2 genome is completely methylated at all 5′-CG-3′ sequences as determined by genomic sequencing. Exactly the same nucleotide sequence is not methylated at all in another Ad2-transformed hamster cell line, HE2 (Toth et al. 1989, 1990). In HE1, the E2A gene has been silenced; in HE2, it is transcribed and translated into the DNA binding protein (Johannsson et al. 1978).
4. The large segment of human cellular DNA sequences in the symmetric Ad12 DNA recombinant SYREC2 (Deuring et al. 1981; Deuring and Doerfler 1983) is not methylated in its 5′-CCGG-3′ sequences in the SYREC2 genome isolated from purified virions. The same cellular nucleotide sequences are, however, very heavily methylated in the 5′-CCGG-3′ sequences inside the human cellular genome in cells growing in culture.

I, therefore, tentatively conclude that the de novo methylation mechanism is not predominantly regulated simply by a specific nucleotide sequence, but that additional parameters such as location in different intranuclear compartments, DNA structure, the type of proteins bound at such structures, the site of foreign DNA integration and the replicative state of the cell may also have an important influence.

Next to nothing, unfortunately, is known about the enzymatic mechanism of de novo methylation. It is not clear whether de novo and maintenance methylations are effected by the same, by different, or by one enzyme in conjunction with different cofactors. It cannot yet be decided how many DNA methyltransferases exist in mammalian cells (Li et al. 1993).

In our own work on DNA methyltransferases, we turned to studies on frog virus 3 (FV3), a member of the iridovirus group (Willis et al. 1989). The FV3 genome in the virion is heavily methylated (Willis and Granoff 1980). Using the genomic sequencing method, we could demonstrate that, in the viral DNA segments investigated, all 5′-CG-3′ sequences are methylated and that 5-mC occurs exclusively in these dinucleotide sequences (SCHETTER et al. 1993). The use of several methylation-sensitive restriction enzymes has demonstrated it to be likely that all 5'-CG-3' sequences in the entire genome are methylated (WILLIS and GRANOFF 1980; SCHETTER et al. 1993). There is evidence that after infection of cells with FV3 the newly synthesized viral DNA is not methylated immediately after replication, but becomes rapidly and completely de novo methylated shortly thereafter. This system thus offers the possibility to study characteristics of the enzyme(s) involved in this de novo methylation. Such studies have been initiated (SCHETTER et al. 1993).

3.3 Methylation of Triplet Repeat Amplifications in the Human Genome: Manifestation of the Cellular Defense Mechanism?

In human genetic diseases, apparently autonomous amplifications of naturally present triplet repeats in the human genome have been recognized to be associated with diseases such as myotonic dystrophy, the fragile X syndrome (FRAXA), Kennedy disease, Huntington's disease, mental retardation found with the fragile site FRAXE on the human X chromosome, spinocerebellar ataxia type I, or hereditary dentatorubro-pallidolysian ataxia (CASKEY et al. 1992; RICHARDS and SUTHERLAND 1992; RIGGINS et al. 1992; THE HUNTINGTON DISEASE COLLABORATION RESEARCH GROUP 1993; KNIGHT et al. 1993; ORR et al. 1993; KOIDE et al. 1994; NAGAFUCHI et al. 1994). These amplifications can lie in the coding sequence of genes or in their 3' or 5' located noncoding regions. It is not understood in detail how this new class of mutations can be linked to the pathogenetic mechanisms of these ailments which frequently involve the central nervous system. Moreover, the (enzymatic) mechanism by which the amplifications are generated is still a puzzle. The possibility furthermore exists that the mutations causing these amplifications lie in genes remote from the sites of triplet repeats, possibly in gene(s) involved in the regulation of DNA replication.

We have observed that synthetic oligodeoxyribonucleotides, like $(CGG)_{17}$, $(GCC)_{17}$, $(CG)_{25}$, but not $(TAA)_{17}$, $(CAGG)_{13}$, by themselves can be in vitro expanded by Taq polymerase under the conditions of polymerase chain reaction (PCR) or by Klenow polymerase (without cycling) to chains of up to 2000 bp. This in vitro amplification, which apparently requires a certain nucleotide sequence and a specific structure dependent on it, can be inhibited, though not obliterated, by the methylation of the C residues in the oligodeoxyribonucleotides, in that much shorter chains are then synthesized in vitro (BEHN-KRAPPA and DOERFLER 1994).

In some of the amplified sequences in human genetic diseases, such as in the FMR-1 gene in the fragile X syndrome (OBERLÉ et al. 1991), an increase in DNA methylation in these sequences has been observed. We have suggested (BEHN-KRAPPA and DOERFLER 1994) that these de novo methylations may represent another manifestation of the cellular defense mechanism against foreign DNA mentioned above. According to this reasoning, the amplified triplet repeats amounting in some cases to more than 2000 nucleotide pairs in excess of the original sequence at this site with many 5'-CG-3' dinucleotides could be recognized as foreign by the DNA methyltransferase system of the host cell, and could thus become de novo methylated, perhaps as a shield against further expansion. This prediction is consistent with the results of our in vitro amplification studies and the effect of 5'-CG-3' methylation which inhibits amplification.

4 Alterations in Patterns of Cellular DNA Methylation and Gene Expression

We investigated the possibility that the insertion of foreign DNA into an established mammalian genome can lead to far-reaching alterations in patterns of cellular DNA methylation and gene expression. These alterations might contribute to the oncogenic transformation of cells at least as significantly as some of the viral gene products thought to be involved in the process.

In the pursuit of this concept, we were initially able to demonstrate that the methylation state in hamster cell DNA sequences immediately abutting the site of insertion of Ad12 DNA in the Ad12-induced tumor T1111/2 have been altered: all the 5'-CG-3' sequences that are completely methylated in normal hamster DNA from animals or from cells devoid of adenovirus DNA have lost the 5-mC residues in the tumor DNA (LICHTENBERG et al. 1987, 1988). This investigation was limited to a sequence of about 1000 bp of cellular DNA.

More recently, we screened the hamster cell genome with several different, randomly selected genomic DNA or cDNA probes by Southern blot hybridization after cutting the hamster cell DNA with HpaII, MspI, or HhaI. Cellular DNA was extracted from a number of Ad12-transformed hamster cell lines and Ad12-induced tumors, from normal BHK21 hamster cells, and from Ad12-infected BHK21 hamster cells at 30 h or 27 days post infection as controls. Among the different cellular hybridization probes, several showed very striking increases in DNA methylation in cellular genes in some of the Ad12-transformed cells and in some of the Ad12-induced tumor cells, while others showed no changes (HELLER et al. 1995). It was shown by fluorescent in situ hybridization (LICHTER and CREMER 1992) that the hybridization probes used in these experiments were located on many different hamster chromosomes which were different from the chromosome carrying the Ad12 genomes (HELLER et al. 1995). One of the cellular DNA hybridization probes used in these studies, the intracisternal A particle (IAP) DNA, has been shown to be represented extensively on many different hamster chromosomes. The IAP DNA exhibited a very striking increase in DNA methylation in cell line T637 as compared with BHK21 hamster cells. IAP DNA, an endogenous retrotransposon-like sequence, is represented about 1000-fold in the hamster genome (LUEDERS and KUFF 1981).

In order to distinguish between the alternative explanations, namely, that these changes in patterns of cellular gene methylation are due to the insertion of foreign DNA into the established hamster cell genome and due to transformation or to the action of early Ad12 gene products synthesized in Ad12-transformed or Ad12-induced tumor cells, we also analyzed cellular DNA isolated from BHK21 cells at 30 h or 27 days after infection with Ad12. In these cells, none of the aforementioned alterations in cellular DNA methylation were observed. We, therefore, favor the interpretation that it seems to be the insertion of the Ad12 genomes into the established hamster genome in conjunction with viral transformation that somehow elicit these far-reaching changes. The mechanism, by which these changes are effected, is not known. Since the patterns of early Ad12 DNA expression are similar in Ad12-infected hamster cells and in Ad12-transformed hamster cells, it is unlikely that Ad12 gene products play the decisive role in rapidly changing the patterns of methylation in cellular genes of transformed cells. It should be emphasized that there are considerable differences in the extent and locations of these changes in different Ad12-transformed cells or Ad12-induced hamster tumor cells. Moreover, only a subset of genes seems to be affected. The cellular genes with altered methylation patterns are located on hamster chromosomes that are definitely different from the chromosome on which Ad12 DNA is integrated. Since we have used only a relatively small number of randomly selected hamster gene probes and found changes in DNA methylation in a high proportion of them, these alterations might be widely distributed.

It has been shown in many different systems that changes in DNA methylation are associated with changes in patterns of gene expression. In Ad12-transformed cells, we demonstrated previously that among 40 different genes tested, the expression of 5 genes was altered in comparison with non-Ad12-transformed BHK21 cells (Rosahl and Doerfler 1992). Although the ratio of the number of genes with alterations in expression to the number of genes tested argues for frequent changes, more work will be required to support these interpretations. We continue to pursue the possibility that the integration of foreign DNA into the hamster genome is associated with widespread changes in DNA methylation and consequently in expression patterns among hamster cellular genes. These findings have consequences for the interpretation of viral oncology, for gene therapy and for the interpretation of results obtained with transgenic organisms.

5 Uptake of Foreign DNA Through the Gastrointestinal Tract

We set out to explore the possibility that traces of foreign DNA that is constantly ingested with the routine food intake might be taken up by the cells of an organism and become integrated at random into the cellular genome. By subsequently eliciting alterations in the methylation and expression patterns of the affected cells, targets are constantly generated in which these changes might lead to the oncogenic transformation of individual cells. This—as yet hypothetical—mechanism of oncogenic transformation is related to the very real daily exposure of the animals' gastrointestinal tract to foreign DNA.

We investigated whether foreign DNA taken up by mammals with the food supply can, at least in part, survive the digestive regime of the gastrointestinal tract and eventually enter into the bloodstream (SCHUBBERT et al. 1994). In model experiments, 3- to 6-month-old mice were fed with bacteriophage M13 DNA (HOFSCHNEIDER 1963) in amounts between 10 and 50 μg. The DNA was supplied in the double-stranded supercoiled circular or EcoRI-linearized form directly by pipette-feeding to the animals' oral cavity or to the food pellets. M13 DNA was chosen as a traceable food additive because we failed to find any homologies between this phage DNA and the DNA repurified from the feces of control mice that had never received this DNA. Moreover, the entire nucleotide sequence of this viral DNA had been determined (YANISCH-PERRON et al. 1985). At various times after feeding M13 DNA, DNA was extracted: (a) from the feces, either extracorporeally or taken from the animals rectum; or (b) from whole blood, from isolated blood cells or from the serum.

These DNA preparations were subsequently analyzed for the presence of M13 DNA sequences by electrophoresis and SOUTHERN (1975) blot hybridization, by dot blot hybridization, or by PCR (SAIKI et al. 1988).

The results of these analyses (SCHUBBERT et al. 1994) demonstrate that:

1. M13 DNA sequences can be detected in the animals feces between 1 and 7 h after feeding
2. M13 DNA sequences are present in the bloodstream 2–4 h after feeding
3. The M13 DNA in blood is exclusively found in the white blood cells (SCHUBBERT and DOERFLER 1994).

In the feces, about 2%–4% of the ingested DNA can be recovered, while in the bloodstream the recoverable percentage lies between 0.01% and 0.1%. The bulk of the feces-excreted M13 DNA is in the size range between 100 and 400 nucleotide pairs, but when PCR has been used, M13 DNA molecules up to a length of 1692 nucleotide pairs are discovered. The PCR-amplified M13 DNA has been resequenced and, apart from occasional, nonsystematic deviations, has been found to be identical with the published nucleotide sequence. The results of these studies have been identical, regardless of whether DNA has been extracted from extracorporeally deposited feces or from feces removed from the terminal gut of the animals. The latter precaution precludes the possibility that the feces extracorporeally deposited might have been externally contaminated by unnoticed oral contacts with the animals.

Similarly and surprisingly, M13 DNA sequences were also detected, both in DNA extracted from total blood and particularly from white blood cells of M13-fed animals. The maximal lengths of M13 DNA fragments observed were 976 nucleotide pairs. Upon resequencing this PCR-amplified DNA, authentic M13 DNA was found with rare, nonsystematic deviations in nucleotide sequence. DNA isolated from the bloodstream of animals that had not been fed M13 DNA was consistently found to be free of M13 DNA by any of the analytical techniques applied.

These data have been confirmed with DNA preparations from 50 different animals (fecal samples) plus 16 buffer-fed (0.01 Tris-HCl, pH 7.5; 1 m*M* EDTA) controls, and from 105 different animals (blood samples) plus 30 buffer-fed controls with essentially identical results (Schubbert et al. 1994). We are currently investigating whether integrated M13 DNA sequences can be cloned by suitable vector systems from the DNA of white blood cells taken from M13-fed animals.

The results described (Schubbert et al. 1994) verify that foreign DNA ingested with food is not completely degraded in the gastrointestinal tract and can reach the bloodstream, although in minute amounts and in fragmented form. Of course, it is known that fragments of DNA are highly recombinogenic. Since this exposure of most, if not all, organ systems to recombinogenic DNA fragments is continuous over the entire life span of an organism, it will be very interesting to consider their contributions to oncogenic, mutagenic, or other pathogenic events which are cumulative over the duration of the individuals life span. Evolutionary implications could only be speculated on.

6 A Concept for Oncogenesis: Implications for Gene Therapy and Research on Transgenic Organisms

The classical concept of insertional mutagenesis relates to damage caused to cellular functions or genes that are encoded at the sites of foreign (adenoviral) DNA insertion into the host genome. Since a very sizeable part of the mammalian genome consists of repetitive sequences with essentially unknown functions, insertion might frequently be nonconsequential to the repertoire of cellular functions. However, we have adduced evidence that the insertion of adenovirus DNA into the hamster cell genome, possibly in conjunction with the transformed cellular phenotype, can be associated with extensive changes in the methylation patterns of cellular genes and of their expression. In that way, foreign DNA insertion at a restricted region on one chromosome could have important sequelae for the expression profile of the afflicted cell, involving a large but so far unknown number of remote cellular locations. It is likely that each insertion event generates a different pattern of changes and, in that sense, a unique disturbance in the recipient nucleus. The overall consequences for cellular survival probably span a wide continuum from cell death to the absence of detectable functional changes. For oncogenic transformation to ensue with derailed growth control, very specific subsets of alterations in methylation and expression patterns may be required. These considerations about possible consequences of foreign DNA insertion broaden the conventional concept of insertional mutagenesis in that, by altering cellular DNA methylation, many parts of the cellular genome could be affected that are located remote from the site of foreign DNA insertion.

Does insertion of foreign DNA into an established genome elicit a signal to the DNA methyltransferase system of the cell? How could such a signal be transmitted to distant parts of the genome involving DNA on different chromosomes? Could the nuclear matrix play a part in the transmission events? Obviously, we do not yet understand these mechanisms, and a great deal of experimental work will be required to research these and other possibilities.

By linking the observation of apparently frequent alterations in patterns of DNA methylation in cellular genes upon insertion of and transformation by the adenovirus genome to the discovery that food-ingested DNA does reach the bloodstream of mammals with the potential of dissemination to many organs of the animal, a concept for oncogenic transformation arises. How frequently do food-ingested DNA fragments obtain access to cells of the organism and become integrated into their genome? How specific or variable are the changes in DNA methylation and expression patterns and do they cause cell transformation or various stages of loss of growth control? One can imagine a very wide gamut of possibilities that will be difficult to prove or disprove in an individual tumor incidence in which it will be impossible to differentiate between primary and secondary events, e.g., changes in DNA methylation and expression patterns, which may be consequences rather than the causes of oncogenesis. This dilemma has accompanied decades of research in many fields of tumor biology.

In the future, we will extend research to the basic mechanisms of foreign DNA insertion in animals, to its frequency and sites of insertion, and to changes in patterns of DNA methylation. For this latter aspect, it would be advisable to use foreign DNA as a model that does not have coding capacity or cannot be expressed in eukaryotic cells. This approach would circumvent the difficulty of having to differentiate between the effects of the insertion event and of gene products of biologically active DNA, such as Ad12 DNA. A wide field of research will have to be addressed here. These projects will be relevant for oncogenesis as well as for gene therapy and the interpretation of experiments in which transgenic animals or plants are utilized. In gene therapy and in transgenic organisms, foreign DNA may affect and alter many parts of an established genome at sites quite distant from the targeted site of insertion via changes in DNA methylation. Although these ideas will complicate the interpretation of some experiments, these concepts will be carefully weighed in the design of future projects.

Acknowledgements. I am indebted to Petra Böhm for expert editorial work. Research in the author's laboratory was supported by the Deutsche Forschungsgemeinschaft (SFB274-A1), the Bundesministerium for Forschung und Technologie (project 2.03) (Genzentrum Köln), and the Fritz Thyssen-Stiftung.

References

Behn-Krappa A, Doerfler W (1994) Enzymatic amplification of synthetic oligodeoxyribonucleotides: implications for triplet repeat expansions in the human genome. Hum Mutat 3: 19–24

Behn-Krappa A, Hölker I, Sandaradura de Silva U, Doerfler W (1991) Patterns of DNA methylation are indistinguishable in different individuals over a wide range of human DNA sequences. Genomics 11: 1–7

Caskey CT, Pizzuti A, Fu Y-H, Fenwick RG Jr, Nelson DL (1992) Triplet repeat mutations in human disease. Science 256: 784–789
Cook JL, Lewis AM Jr (1979) Host response to adenovirus 2-transformed hamster embryo cells. Cancer Res 39: 1455–1461
Deuring R, Doerfler W (1983) Proof of recombination between viral and cellular genomes in human KB cells productively infected by adenovirus type 12: structure of the junction site in a symmetric recombinant (SYREC). Gene 26: 283–289
Deuring R, Klotz G, Doerfler W (1981) An unusual symmetric recombinant between adenovirus type 12 DNA and human cell DNA. Proc Natl Acad Sci USA 78: 3142–3146
Doerfler W (1968) The fate of the DNA of adenovirus type 12 in baby hamster kidney cells. Proc Natl Acad Sci USA 60: 636–643
Doerfler W (1970) Integration of the deoxyribonucleic acid of adenovirus type 12 into the deoxyribonucleic acid of baby hamster kidney cells. J Virol 6: 652–666
Doerfler W (1981) DNA methylation—a regulatory signal in eukaryotic gene expression. J Gen Virol 57: 1–20
Doerfler W (1983) DNA methylation and gene activity. Annu Rev Biochem 52: 93–124
Doerfler W (1984) DNA methylation: role in viral transformation and persistence. Adv Viral Oncol 4: 217–247
Doerfler W (1989) Complexities in gene regulation by promoter methylation. Nucleic Acids Mol Biol 3: 92–119
Doerfler W (1991) Patterns of DNA methylation-evolutionary vestiges of foreign DNA inactivation as a host defense mechanism—a proposal. Biol Chem Hoppe-Seyler 372: 557–564
Doerfler W (1992) Transformation of cells by adenoviruses: less frequently discussed mechanisms. In: Doerfler W, Böhm P (eds) Malignant transformation by DNA viruses. Molecular mechanisms. VCH, Weinheim, Germany, pp 87–109
Doerfler W (1993) Adenoviral DNA integration and changes in DNA methylation patterns: a different view of insertional mutagenesis. Prog Nucleic Acid Res Mol Biol 46: 1–36
Doerfler W, Tatzelt J, Orend G, Schorr J, Rosahl T, Fechteler K, Zock C, Lichtenberg U (1993) Viral and cellular strategies on integrated adenovirus genomes. In: Doerfler W, Böhm P (eds) Virus strategies. VCH, Weinheim, Germany, pp 369–400
Eick D, Stabel S, Doerfler W (1980) Revertants of adenovirus type 12-transformed hamster cell line T637 as tools in the analysis of integration patterns. J Virol 36: 41–49
Ernberg I, Falk K, Minarovits J, Busson P, Tursz T, Masucci MG, Klein G (1989) The role of methylation in the phenotype-dependent modulation of Epstein-Barr nuclear antigen 2 and latent membrane protein genes in cells latently infected with Epstein-Barr virus. J Gen Virol 70: 2989–3002
Groneberg J, Sutter D, Soboll H, Doerfler W (1978) Morphological revertants of adenovirus type 12-transformed hamster cells. J Gen Virol 40: 635–645
Günthert U, Schweiger M, Stupp M, Doerfler W (1976) DNA methylation in adenovirus, adenovirus-transformed cells, and host cells. Proc Natl Acad Sci USA 73: 3923–3927
Hasse A, Schulz WA (1994) Enhancement of reporter gene de novo methylation by DNA fragments from the α-fetoprotein control region. J Biol Chem 269: 1821–1826
Heller H, Wilgenbus P, Doerfler W (1995) The chromosomal insertion of adenovirus type 12 DNA and enhanced methylation of cellular genes in transformed and tumor cells (submitted)
Hofschneider PH (1963) Untersuchungen über kleine E.coli K12 Bakteriophagen. Z Naturforsch 18b: 203–210
Hu, L-F, Minarovits J, Cao, S-L, Contreras-Salazar B, Rymo L, Falk K, Klein G, Ernberg I. (1991) Variable expression of latent membrane protein in nasopharyngeal carcinoma can be related to methylation status of the Epstein-Barr virus BNLF-1 5'-flanking region. J Virol 65: 1558–1567
Johansson K, Persson H, Lewis AM, Pettersson U, Tibbetts C, Philipson L (1978) viral DNA sequences and gene products in hamster cells transformed by adenovirus type 2. J Virol 27: 628–639
Kämmer C, Doerfler W (1995) Genomic sequencing reveals absence of DNA methylation in the major late promoter of adenovirus type 2 DNA in the virion and in productivity infected cells. (submitted)
Knight SJL, Flannery AV, Hirst MC, Campbell L, Christodoulou Z, Phelps SR, Pointon J, Middleton-price HR, Barnicoat A, Pembrey ME, Holland J, Oostra BA, Bobrow M, Davies KE (1993) Trinucleotide repeat amplification and hypermethylation of a CpG island in FRAXE mental retardation. Cell 74: 127–134
Kochanek S, Toth M, Dehmel A, Renz D, Doerfler W (1990) Interindividual concordance of methylation profiles in human genes for tumor necrosis factors alpha and beta. Proc Natl Acad Sci USA 87: 8830–8834

Kochanek S, Renz D, Doerfler W (1993) DNA methylation in the Alu sequences of diploid and haploid primary human cells: functional implication. EMBO J 12: 1141–1151
Koide R, Ikeuchi T, Onodera O, Tanaka H, Igarashi S, Endo K,Takahashi H, Kondo R, Ishikawa A, Hayashi T, Saito M, Tomoda A, Miike T, Naito H, Ikuta F, Tsuji S (1994) Unstable expansion of CAG repeat in hereditary dentatorubral-pallidoluysian atrophy (DRPLA) Nature Genet 6: 9–13
Kruczek I, Doerfler W (1982) The unmethylated state of the promoter/leader and 5'-regions of integrated adenovirus genes correlates with gene expression. EMBO J 1: 409–414
Kuhlmann I, Doerfler W (1982) Shifts in the extent and patterns of DNA methylation upon explantation and subcultivation of adenovirus type 12-induced hamster tumor cells. Virology 118: 169–180
Lettmann C, Schmitz B, Doerfler W (1991) Persistence or loss of preimposed methylation patterns and de novo methylation of foreign DNA integrated in transgenic mice. Nucleic Acids Res 19: 7131–7137
Li E, Beard C, Jaenisch R (1993) Role of DNA methylation in genomic imprinting. Nature 366: 362–365
Lichtenberg U, Zock C, Doerfler W (1987) Insertion of adenovirus type 12 DNA in the vicinity of an intracisternal A particle genome in Syrian hamster tumor cells. J Virol 61: 2719–2726
Lichtenberg U, Zock C, Doerfler W (1988) Integration of foreign DNA into mammalian genome can be associated with hypomethylation at site of insertion. Virus Res 11: 335–342
Lichter P, Cremer T (1992) Chromosome analysis by nonisotopic in situ hybridization. In: Rooney DE, Czepulkowski BH (eds) Human cytogenetics. A practical approach, vol I, Oxford University Press, Oxford, pp 157–192
Linn F, Heidmann I, Saedler H, Meyer P (1990) Epigenetic changes in the expression of the maize A1 gene in Petunia hybrida: role of numbers of integrated gene copies and state of methylation. Mol Gen Genet 222: 329–336
Lueders KK, Kuff EL (1981) Sequences homologous to retrovirus-like genes of the mouse are present in multiple copies in the Syrian hamster genome. Nucl Acids Res 9: 5917–5930
Muiznieks I, Doerfler W (1994a) The impact of 5'-CG-3' methylation on the activity of different eukaryotic promoters: a comparative study. FEBS Lett 334: 251–254
Muiznieks I, Doerfler W (1994b) The topology of the promoter of RNA polymerase II- and III- transcribed genes is modified by the methylation of 5'-CG-3' dinucleotides. Nucl Acids Res 22: 2568–2575
Mummaneni P, Bishop PL, turker MS (1993) A cis-acting element accounts for a conserved methylation pattern upstream of the mouse adenine phosphoribosyl transferase gene. J Biol Chem 268: 552–558
Nagafuchi S, Yanagisawa H, Sato K, Shirayama T, Ohsaki E, Bundo M, Takeda T, Tadokoro K, Kondo I, Murayama N, Tanaka Y, Kikushima H, Umino K, Kurosawa H, Furukawa T, Nihei K, Inoue T, Sano A, Komure O, Takahashi M, Yoshizawa T, Kanazawa I, Yamada M (1994) Dentatorubal and pallidoluysian atrophy expansion of an unstable CAG trinucleotide on chromosome 12p. Nature Genet 6: 14–18
Oberlé I, Rousseau F, Heitz D, Kretz C, Devys D, Hanauer A, Boué J, Bertheas MF, Mandel JL (1991) Instability of a 550-base pair DNA segment and abnormal methylation in fragile X syndrome. Science 252: 1097–1102
Orend G, Kuhlmann I, Doerfler W (1991) Spreading of DNA methylation accross integrated foreign (adenovirus type 12) genomes in mammalian cells. J Virol 65: 4301–4308
Orend G, Knoblanch M, Kämmer C, Tjia ST, Schmitz B, Linkwitz A, Meyer zu Altenschildesche G, Maas J, Doerfler W (1995) The initiation of de novo methylation of foreign DNA integrated into a mammlian genome is not exclusively targeted by nucleotide sequence. J. Virol 69 (in press)
Orr HT, Chung M-Y, Banfi S, Kwiatkowski TJ Jr, Servadio A, Beaudet AL, McCall AE, Duvick LA, Ranum LP, Zoghbi HY (1993) Expansion of an unstable trinucleotide CAG repeat in spinocerebellar ataxia type 1. Nature Genet 4: 221–226
Ragot T, Vincent N, Chafey P, Vigne E, Gilgenkrantz H, Couton D, Cartaud J, Briand P, Kaplan J-C, Perricaudet M, Kahn A (1993) Efficient adenovirus-mediated transfer of a human minidystrophin gene to skeletal muscle of mdx mice. Nature 361: 647–650
Reik W, Collick A, Norris ML, Barton SC, Surani MA (1987) Genomic imprinting determines methylation of parental alleles in transgenic mice. Nature 328: 248–251
Richards RI, Sutherland GR (1992) Dynamic mutations: a new class of mutations causing human disease. Cell 70: 709–712
Riggins GJ, Lokey LK, Chastain JL, Leiner HA, Sherman SL, Wilkinson KD, Warren ST (1992) Human genes containing polymorphic trinucleotide repeats. Nature Genet 2: 186–191
Rosahl T, Doerfler W (1992) Alterations in the levels of expression of specific cellular genes in adenovirus-infected and -transformed cells. Virus Res 26: 71–90
Saiki RK, Gelfand DH, Stoffel S, Scharf SJ, Higuchi R, Horn GT, Mullis KB, Erlich HA (1988) Primer-

directed enzymatic amplification of DNA with a thermostable DNA polymerase. Science 239: 487–491

Sapienza C, Peterson AC, Rossant J, Balling R (1987) Degree of methylation of transgenes is dependent on gamete of origin. Nature 328: 251–254

Schetter C, Grünemann B, Hölker I, Doerfler W (1993) Patterns of frog virus 3 DNA methylation and DNA methyltransferase activity from nuclei of infected cells. J Virol 67: 6973–6978

Schröer J, Doerfler W (1995) Extensive association of adenovirus type 12 DNA with chromosomes early after the infection of hamster and human cells. (submitted)

Schubbert R, Doerfler W (1995) Fragments of foreign DNA ingested by mice can be recovered in peripheral white blood cells. (submitted)

Schubbert R, Lettmann CM, Doerfler W (1994) Ingested foreign DNA (phage M13) survives transiently in the gastrointestinal tract and enters the bloodstream of mice. Mol Gen Genetics 242: 495–504

Southern EM (1975) Detection of specific sequences among DNA fragments separated by gel electrophoresis. J Mol Biol 98: 503–517

Surani MAH, Barton SC, Norris ML (1984) Development of reconstituted mouse eggs suggests imprinting of the genome during gametogenesis. Nature 308: 548–550

Sutter D, Doerfler W (1979) Methylation of integrated viral DNA sequences in hamster cells transformed by adenovirus 12. Cold Spring Harb Symp Quant Biol 44: 565–568

Sutter D, Doerfler W (1980) Methylation of integrated adenovirus type 12 DNA sequences in transformed cells is inversely correlated with viral gene expression. Proc Natl Acad Sci USA 77: 253–256

Sutter D, Westphal M, Doerfler W (1978) Patterns of integration of viral DNA sequences in the genomes of adenovirus type 12-transformed hamster cells. Cell 14: 569–585

Swain JL, Stewart TA, Leder P (1987) Parental legacy determines methylation and expression of an autosomal transgene: a molecular mechanism for parental imprinting. Cell 50: 719–727

Szyf M, Tanigawa G, McCarthy PL Jr (1990) A DNA signal from the Thy-1 gene defines de novo methylation patterns in embryonic stem cells. Mol Cell Biol 10: 4396–4400

The Huntington's Disease Collaborative Research Group (1993) A novel gene containing a trinucleotide repeat that is expanded and unstable on Huntington's disease chromosomes. Cell 72: 971–983

Toth M, Lichtenberg U, Doerfler W (1989) Genomic sequencing reveals a 5-methyldeoxycytosine-free domain in active promoters and the spreading of preimposed methylation patterns. Proc Natl Acad Sci USA 86: 3728–3732

Toth M, Müller U, Doerfler W (1990) Establishment of de novo DNA methylation patterns. Transcription factor binding and deoxycytidine methylation at CpG and non-CpG sequences in an integrated adenovirus promoter. J Mol Biol 214: 673–683

Vardimon L, Neumann R, Kuhlmann I, Sutter D, Doerfler W (1980) DNA methylation and viral gene expression in adenovirus-transformed and -infected cells. Nucleic Acids Res 8: 2461–2473

Wienhues U, Doerfler W (1985) Lack of evidence for methylation of parental and newly synthesized adenovirus type 2 DNA in productive infections. J Virol 56: 320–324

Willis DB, Granoff A (1980) Frog virus 3 DNA is heavily methylated at CpG sequences. Virology 107: 250–257

Willis DB, Thompson JP, Essani K, Goorha R (1989) Transcription of methylated viral DNA by eukaryotic RNA polymerase II. Cell Biophysics 15: 97–111

Yanisch-Perron C, Vieira J, Messing J (1985) Improved M13 phage cloning vectors and host strains: nucleotide sequences of the M13mp18 and pUC19 vectors. Gene 33: 103–119

Subject Index

Current Topics in Microbiology and Immunology

Volumes published since 1989 (and still available)

Vol. 152: **Bosma, Melvin J.; Phillips, Robert A.; Schuler, Walter (Ed.):** The Scid Mouse. Characterization and Potential Uses. EMBO Workshop held at the Basel Institute for Immunology, Basel, Switzerland, February 20–22, 1989. 1989. 72 figs. XII, 263 pp. ISBN 3-540-51512-7

Vol. 153: **Lambris, John D. (Ed.):** The Third Component of Complement. Chemistry and Biology. 1989. 38 figs. X, 251 pp. ISBN 3-540-51513-5

Vol. 154: **McDougall, James K. (Ed.):** Cytomegaloviruses. 1990. 58 figs. IX, 286 pp. ISBN 3-540-51514-3

Vol. 155: **Kaufmann, Stefan H. E. (Ed.):** T-Cell Paradigms in Parasitic and Bacterial Infections. 1990. 24 figs. IX, 162 pp. ISBN 3-540-51515-1

Vol. 156: **Dyrberg, Thomas (Ed.):** The Role of Viruses and the Immune System in Diabetes Mellitus. 1990. 15 figs. XI, 142 pp. ISBN 3-540-51918-1

Vol. 157: **Swanstrom, Roland; Vogt, Peter K. (Ed.):** Retroviruses. Strategies of Replication. 1990. 40 figs. XII, 260 pp. ISBN 3-540-51895-9

Vol. 158: **Muzyczka, Nicholas (Ed.):** Viral Expression Vectors. 1992. 20 figs. IX, 176 pp. ISBN 3-540-52431-2

Vol. 159: **Gray, David; Sprent, Jonathan (Ed.):** Immunological Memory. 1990. 38 figs. XII, 156 pp. ISBN 3-540-51921-1

Vol. 160: **Oldstone, Michael B. A.; Koprowski, Hilary (Eds.):** Retrovirus Infections of the Nervous System. 1990. 16 figs. XII, 176 pp. ISBN 3-540-51939-4

Vol. 161: **Racaniello, Vincent R. (Ed.):** Picornaviruses. 1990. 12 figs. X, 194 pp. ISBN 3-540-52429-0

Vol. 162: **Roy, Polly; Gorman, Barry M. (Eds.):** Bluetongue Viruses. 1990. 37 figs. X, 200 pp. ISBN 3-540-51922-X

Vol. 163: **Turner, Peter C.; Moyer, Richard W. (Eds.):** Poxviruses. 1990. 23 figs. X, 210 pp. ISBN 3-540-52430-4

Vol. 164: **Bækkeskov, Steinnun; Hansen, Bruno (Eds.):** Human Diabetes. 1990. 9 figs. X, 198 pp. ISBN 3-540-52652-8

Vol. 165: **Bothwell, Mark (Ed.):** Neuronal Growth Factors. 1991. 14 figs. IX, 173 pp. ISBN 3-540-52654-4

Vol. 166: **Potter, Michael; Melchers, Fritz (Eds.):** Mechanisms in B-Cell Neoplasia. 1990. 143 figs. XIX, 380 pp. ISBN 3-540-52886-5

Vol. 167: **Kaufmann, Stefan H. E. (Ed.):** Heat Shock Proteins and Immune Response. 1991. 18 figs. IX, 214 pp. ISBN 3-540-52857-1

Vol. 168: **Mason, William S.; Seeger, Christoph (Eds.):** Hepadnaviruses. Molecular Biology and Pathogenesis. 1991. 21 figs. X, 206 pp. ISBN 3-540-53060-6

Vol. 169: **Kolakofsky, Daniel (Ed.):** Bunyaviridae. 1991. 34 figs. X, 256 pp. ISBN 3-540-53061-4

Vol. 170: **Compans, Richard W. (Ed.):** Protein Traffic in Eukaryotic Cells. Selected Reviews. 1991. 14 figs. X, 186 pp. ISBN 3-540-53631-0

Vol. 171: **Kung, Hsing-Jien; Vogt, Peter K. (Eds.):** Retroviral Insertion and Oncogene Activation. 1991. 18 figs. X, 179 pp. ISBN 3-540-53857-7

Vol. 172: **Chesebro, Bruce W. (Ed.):** Transmissible Spongiform Encephalopathies. 1991. 48 figs. X, 288 pp. ISBN 3-540-53883-6

Vol. 173: **Pfeffer, Klaus; Heeg, Klaus; Wagner, Hermann; Riethmüller, Gert (Eds.):** Function and Specificity of γ/δ TCells. 1991. 41 figs. XII, 296 pp. ISBN 3-540-53781-3

Vol. 174: **Fleischer, Bernhard; Sjögren, Hans Olov (Eds.):** Superantigens. 1991. 13 figs. IX, 137 pp. ISBN 3-540-54205-1

Vol. 175: **Aktories, Klaus (Ed.):** ADP-Ribosylating Toxins. 1992. 23 figs. IX, 148 pp. ISBN 3-540-54598-0

Vol. 176: **Holland, John J. (Ed.):** Genetic Diversity of RNA Viruses. 1992. 34 figs. IX, 226 pp. ISBN 3-540-54652-9

Vol. 177: **Müller-Sieburg, Christa; Torok-Storb, Beverly; Visser, Jan; Storb, Rainer (Eds.):** Hematopoietic Stem Cells. 1992. 18 figs. XIII, 143 pp. ISBN 3-540-54531-X

Vol. 178: **Parker, Charles J. (Ed.):** Membrane Defenses Against Attack by Complement and Perforins. 1992. 26 figs. VIII, 188 pp. ISBN 3-540-54653-7

Vol. 179: **Rouse, Barry T. (Ed.):** Herpes Simplex Virus. 1992. 9 figs. X, 180 pp. ISBN 3-540-55066-6

Vol. 180: **Sansonetti, P. J. (Ed.):** Pathogenesis of Shigellosis. 1992. 15 figs. X, 143 pp. ISBN 3-540-55058-5

Vol. 181: **Russell, Stephen W.; Gordon, Siamon (Eds.):** Macrophage Biology and Activation. 1992. 42 figs. IX, 299 pp. ISBN 3-540-55293-6

Vol. 182: **Potter, Michael; Melchers, Fritz (Eds.):** Mechanisms in B-Cell Neoplasia. 1992. 188 figs. XX, 499 pp. ISBN 3-540-55658-3

Vol. 183: **Dimmock, Nigel J.:** Neutralization of Animal Viruses. 1993. 10 figs. VII, 149 pp. ISBN 3-540-56030-0

Vol. 184: **Dunon, Dominique; Mackay, Charles R.; Imhof, Beat A. (Eds.):** Adhesion in Leukocyte Homing and Differentiation. 1993. 37 figs. IX, 260 pp. ISBN 3-540-56756-9

Vol. 185: **Ramig, Robert F. (Ed.):** Rotaviruses. 1994. 37 figs. X, 380 pp. ISBN 3-540-56761-5

Vol. 186: **zur Hausen, Harald (Ed.):** Human Pathogenic Papillomaviruses. 1994. 37 figs. XIII, 274 pp. ISBN 3-540-57193-0

Vol. 187: **Rupprecht, Charles E.; Dietzschold, Bernhard; Koprowski, Hilary (Eds.):** Lyssaviruses. 1994. 50 figs. IX, 352 pp. ISBN 3-540-57194-9

Vol. 188: **Letvin, Norman L.; Desrosiers, Ronald C. (Eds.):** Simian Immunodeficiency Virus. 1994. 37 figs. X, 240 pp. ISBN 3-540-57274-0

Vol. 189: **Oldstone, Michael B. A. (Ed.):** Cytotoxic T-Lymphocytes in Human Viral and Malaria Infections. 1994. 37 figs. IX, 210 pp. ISBN 3-540-57259-7

Vol. 190: **Koprowski, Hilary; Lipkin, W. Ian (Eds.):** Borna Disease. 1995. 33 figs. IX, 134 pp. ISBN 3-540-57388-7

Vol. 191: **ter Meulen, Volker; Billeter, Martin A. (Eds.):** Measles Virus. 1995. 23 figs. IX, 196 pp. ISBN 3-540-57389-5

Vol. 192: **Dangl, Jeffrey L. (Ed.):** Bacterial Pathogenesis of Plants and Animals. 1994. 41 figs. IX, 343 pp. ISBN 3-540-57391-7

Vol. 193: **Chen, Irvin S. Y.; Koprowski, Hilary; Srinivasan, Alagarsamy; Vogt, Peter K. (Eds.):** Transacting Functions of Human Retroviruses. 1995. 49 figs. Approx. IX, 240 pp. ISBN 3-540-57901-X

Vol. 194: **Potter, Michael; Melchers, Fritz (Eds.):** Mechanisms in B-cell Neoplasia. 1995. Approx. 152 figs. XXV, 458 pp. ISBN 3-540-58447-1

Vol. 195: **Montecucco, Cesare (Ed.):** Clostridial Neurotoxins. 1995. Approx. 28 figs. Approx. 260 pp. ISBN 3-540-58452-8

Vol. 196: **Koprowski, Hilary; Maeda Hiroshi (Eds.):** The Role of Nitric Oxide in Physiology and Pathophysiology. 1995. 21 figs. Approx. 100 pp. ISBN 3-540-58214-2

MIX
Papier aus verantwortungsvollen Quellen
Paper from responsible sources
FSC® C105338

If you have any concerns about our products,
you can contact us on
ProductSafety@springernature.com

In case Publisher is established outside the EU,
the EU authorized representative is:
Springer Nature Customer Service Center GmbH
Europaplatz 3, 69115 Heidelberg, Germany

Printed by Libri Plureos GmbH
in Hamburg, Germany